SUPERPAVE

Performance Graded Asphalt Binder Specification and Testing

ASPHALT INSTITUTE

Superpave Series No. 1 (SP-1)
Third Edition, Revised 2003

The Asphalt Institute can accept no responsibility for the inappropriate use of this manual. Engineering judgment and experience must be used to properly utilize the principles and guidelines contained in this manual, taking into account available equipment, local materials and conditions.

All reasonable care has been taken in the preparation of this manual; however, the Asphalt Institute can accept no responsibility for the consequences of any inaccuracies which it may contain.

FOREWORD

The Strategic Highway Research Program (SHRP) was established by Congress in 1987 as a five year, $150 million research program to improve the performance and durability of United States roads and to make those roads safer for both motorists and highway workers. $50 million of the SHRP research funds were used for the development of performance based asphalt specifications to directly relate laboratory analysis with field performance.

Superpave™ (*Su*perior *Per*forming Asphalt *Pave*ments) is a product of the SHRP asphalt research. The Superpave system incorporates performance based asphalt materials characterization with the design environmental conditions to improve performance by controlling rutting, low temperature cracking and fatigue cracking. The three major components of Superpave are the asphalt binder specification, the mixture design and analysis system, and a computer software system. This manual is devoted to the Superpave asphalt binder specification, AASHTO M320, and the tests that it uses to characterize asphalt binders.

Much of the material contained in this manual was originally produced by the Asphalt Institute under contract with the Federal Highway Administration (FHWA) as the National Asphalt Training Center (NATC) for Superpave technology. The NATC developed and conducted week-long training courses in Superpave binder and mixture technology, and was administered by the FHWA's Office of Technology Applications.

A companion publication, ***Superpave Mix Design***, Superpave Series No. 2 (SP-2) is also available from the Asphalt Institute.

Superpave™ is a trademark of the Strategic Highway Research Program.

CONTENTS

List of Figures

List of Tables

CHAPTER 1

Origins of Asphalt

Asphalt is defined by the American Society for Testing and Materials (ASTM) as "a dark brown to black cementitious material in which the predominating constituents are bitumens which occur in nature or are obtained in petroleum processing." As cement, asphalt is especially valuable to the engineer because it is strong, readily adhesive, highly waterproof, and durable. It provides limited flexibility to mixtures of mineral aggregates with which it is usually combined. Although a solid or semi-solid at ordinary atmospheric temperatures, asphalt may be readily liquefied by applying heat, by dissolving it in petroleum solvents, or by emulsifying it in water.

Man has long been aware of the adhesive and waterproofing properties of asphalt. Surface accumulations of petroleum, forced upward by geological forces, leave behind naturally-occurring lakes of asphalt that have hardened after exposure to the elements. Current examples of these deposits include Trinidad Lake Asphalt, on the Island of Trinidad off the northern coast of Venezuela, and the La Brea "Tar" Pits near Los Angeles. Natural asphalt is also found impregnated within porous rock, such as sandstone or limestone, called rock asphalt. Natural asphalts were used by the ancient Babylonians, Egyptians, Greeks, and Romans as a road-building and waterproofing material. Even with the long history of asphalt usage, the evolution of asphalt as a widespread ingredient in paving material did not occur until modern petroleum-refining techniques were developed in the early 1900's.

The first documented use of rock asphalt as a sidewalk surfacing occurred in France in 1802 and later in Philadelphia in 1838. In 1870, the first asphalt pavement was constructed in Newark, New Jersey. The first sheet asphalt (fine sand mix) pavement was built in Washington, D.C. in 1876 with imported lake asphalt. Finally, in 1902, asphalt began to be refined from petroleum, which led to the development of the asphalt paving industry in the US. This rapid growth in asphalt usage is shown in Figure 1.1.

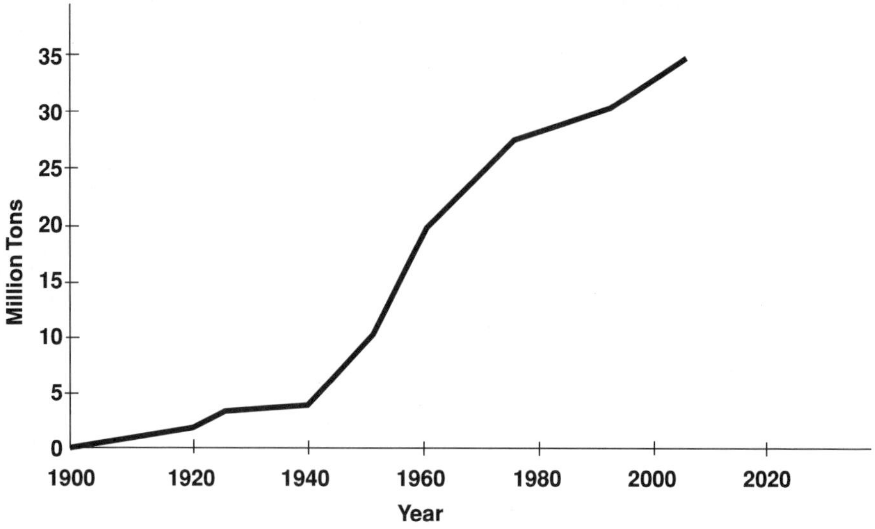

Figure 1.1 Annual Asphalt Usage

Asphalt is also used in hundreds of products not related to traditional construction and transportation industries, as described in *Introduction to Asphalt*, MS-5, Asphalt Institute. However, about 85 percent of the asphalt refined from petroleum (the source of nearly all asphalt today) continues to be used as a paving material, commonly called asphalt cement. Approximately 50 companies in the US and Canada currently produce asphalt cement refined from crude oil. Few of these companies produce asphalt cement as their principal product. In the majority of cases, the asphalt used to pave roads is made from the residuum that remains after the refineries remove the distillates to manufacture gasoline, jet fuel, kerosene, lubricating oils and other commodities.

Before spot market trading of crude oil was common, refiners rarely changed their crude oil source, or feedstock. This practice led to asphalt sources that had consistent properties over long periods of time. Out of the need to react to changes in the world marketplace, modern integrated refineries can now switch crude feedstock blends in a few days. This enhanced refining capability has created more variation in asphalt cement properties, both chemical and physical. New oil exploration in areas such as the Alaska North Slope and the Maya Peninsula in Mexico has added to the variation in asphalt properties. Out of this variation comes a greater need to be able to evaluate asphalt response to particular conditions and predict performance in terms of recognized forms of distress.

Chemical Composition of Asphalt

Asphalt is made from crude petroleum, a product formed naturally from organic matter over millions of years under varying conditions of temperature and pressure. From this beginning, about 90 to 95 percent by weight of asphalt is composed of carbon and hydrogen, and, therefore it is called a hydrocarbon. The remaining portion consists of two types of atoms: heteroatoms and metals. The heteroatoms, such as nitrogen, oxygen, and sulfur, often replace carbon atoms in the asphalt molecular structure. These contribute to many of asphalt's unique chemical and physical properties by causing much of the interaction between molecules. The type and amount of heteroatoms that exist in an asphalt are a function of both the crude source and its exposure to aging. Heteroatoms, especially sulfur, react more easily than carbon and hydrogen to incorporate oxygen (oxidize). Oxidation is the primary part of the overall asphalt aging process; evaporation (volatilization) and degradation associated with light (photodegradation) also contribute.

Metal atoms, such as vanadium, nickel, and iron, are present in trace quantities, typically far less than one percent. The main significance of the metal atoms is that these can provide some indication or "fingerprint" of the asphalt crude source. Table 1.1 (from Petersen, Transportation Research Record 999, 1984) shows an elemental analysis of four representative petroleum asphalts.

Table 1.1 Elemental Analysis of Four Asphalt Cements

Asphalt Cement	A	B	C	D
Carbon, percent	83.77	85.78	82.90	86.77
Hydrogen, percent	9.91	10.19	10.45	10.93
Nitrogen, percent	0.28	0.26	0.78	1.10
Sulfur, percent	5.25	3.41	5.43	0.99
Oxygen, percent	0.77	0.36	0.29	0.20
Vanadium, ppm	180.	7.	1380.	4.
Nickel, ppm	22.	0.4	109.	6.

Asphalt components can be separated and evaluated by using the solubility of the various molecules in different solvents. The two most frequently-used fractionation methods are the chromatographic method of Corbett (ASTM D 4124, *Standard Test Method for Separation of Asphalt into Four Fractions*) and the precipitation method of Rostler (described in Proceedings, Association of Asphalt Paving Technologists, Volume 31, 1962). The generic fractions determined from these methods are complex mixtures with varying properties and are not discreet chemical species, which has led to confusing and contradicting predictions of behavior. The two methods produce different fractions because of the inherent differences in separation. The Rostler method uses sulfuric acid of varying strength to analytically separate the groups while the Corbett method uses differential absorption and desorption. Both methods separate out the insoluble discrete "asphaltenes" in different quantities using two different solvents (n-heptane or n-pentane) as shown in Figure 1.2. The remaining groups of resins and oils are defined differently by each procedure. Corbett refers to this mix as petrolenes, made up of saturates, napthene aromatics, and polar aromatics. Rostler refers to this mix as maltenes, composed of paraffins, first and second acidiffins, and nitrogen bases. The figure shows some vague relationships between these two sets of components; however, overlaps exist and there are no strict dividing lines. Some information and relative tendencies have been observed from these analytical methods, but no definite correlations between component quantities and asphalt performance have been identified.

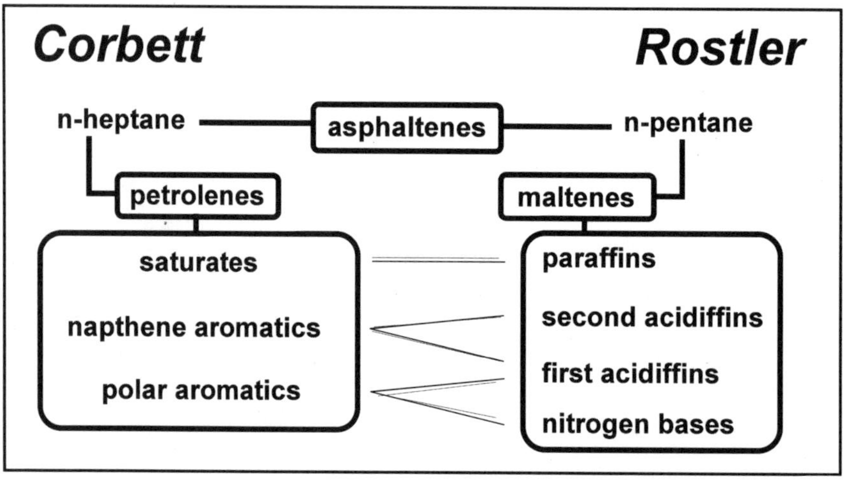

Figure 1.2 Component Analysis of Asphalt Cement

The molecular structure of asphalt is extremely complex, and varies in size and type of chemical bonding with each blend or source. There are three basic types of molecules: aliphatics, cyclics, and aromatics (Figure 1.3). The aliphatics, or paraffinics, are linear, three-dimensional, chain-like molecules and are "oily" or "waxy" in nature. The cyclics, or naphthenics, are three-dimensional saturated rings of carbon atoms with various atoms attached. The aromatics are flat, stable rings of carbon atoms that easily stack together and have a strong odor. All of these types of molecules interact with one another in various ways to affect the physical and chemical behavior of an asphalt.

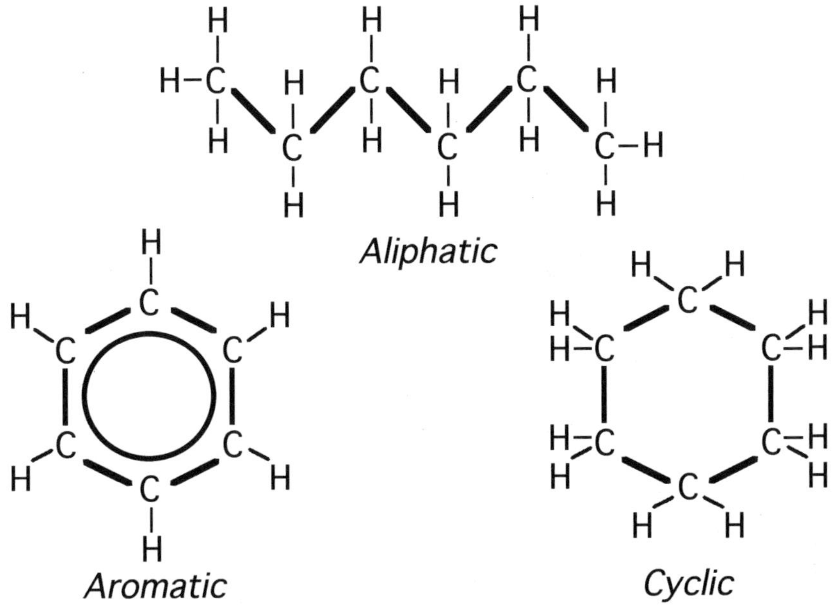

Figure 1.3 Types of Asphalt Molecules

The chemical bonds holding the molecules together are relatively weak and are easily broken by heat or shear stress, which explains the viscoelastic nature of asphalt. For example, the intermolecular bonding is destroyed when asphalt is heated and flows freely. When the asphalt cools, the weak bonds reform and the chemical structure returns, but not necessarily to the same structure as before heating. All of the molecules contained in asphalt fall in one of two functional categories: polar and non-polar. The polar molecules form a "network" and give asphalt its elastic properties. The non-polar molecules form the body of material around the network and contribute to the viscous properties of asphalt. Polar and non-polar molecules exist together in a homogeneous mixture. The weak interactions between molecules result in the behavior of asphalt at elevated temperatures as a Newtonian fluid; the viscosity change is proportional to the amount of temperature change.

The actual formation of the polar network in hot mix asphalt is a function of the aggregate mineral type and the environment (air quality and quantity, pressure, temperature) around the specific asphalt during mixing. Current research indicates that a "proper" balance between polar and non-polar molecules appears critical to good performance. Asphalt with large quantities of high-molecular weight, non-polar molecules tends to exhibit poor low-temperature behavior or brittleness. Other trends in asphalt behavior can be related to or explained by the chemical composition of asphalt; however, the complexity and changing chemical structure of asphalts make it extremely difficult to use chemical analyses to characterize performance. For this reason, physical property measurements continue to be the primary means of specifying and selecting asphalt cement.

Asphalt Behavior

Because of its viscoelastic nature, asphalt cement behavior depends on both temperature and rate of loading. As shown in Figure 1.4, the amount an asphalt flows could be the same for one hour at 60°C or 10 hours at 25°C. In other words, the effects of time and temperature are related; the behavior at high temperature over short time periods is equivalent to what occurs at lower temperatures and long durations. This is often referred to as the time-temperature shift or superposition concept of asphalt cement.

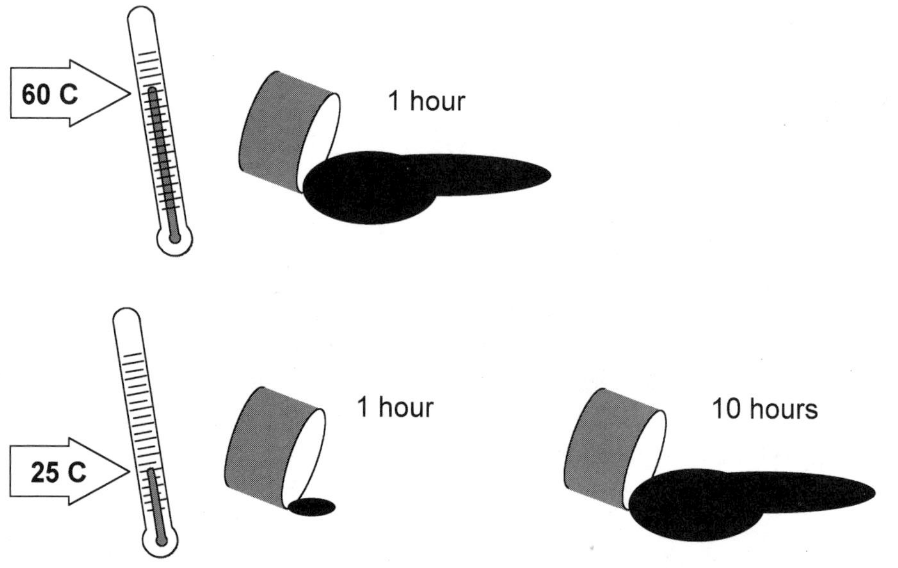

Figure 1.4 Asphalt Cement Flow Behavior

➤➤ *High Temperature Behavior*

In hot conditions (e.g., desert climate) or under sustained loads (e.g., slow moving or parked trucks), asphalt cement acts like a viscous liquid. [Under these circumstances, the aggregate is the part of the hot mix asphalt that bears the load.] Viscosity is the physical material characteristic used to describe the resistance of liquids to flow. If the slow-flow movement of hot asphalt cement could be observed under a powerful microscope, adjacent layers of molecules would be observed sliding past each other (Figure 1.5). The resisting force or friction between these layers is related to the relative velocity at which they slide by each other. The top layer tries to pull the bottom layer along while the bottom layer tries to hold the top layer back. The relationship between the resisting force and relative velocity can be very different for different liquids.

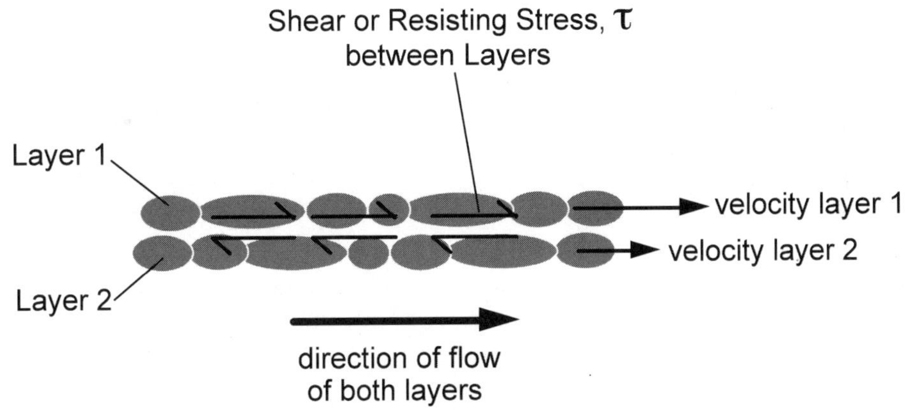

Figure 1.5 Microscopic View of Liquid Flow Characteristics

Fortunately, viscosity is one characteristic (as opposed to chemical composition, for example) that can be used to express this difference. This equation describes the situation in Figure 1.5 and shows how the coefficient of viscosity (μ) is used to explain differences in flow characteristics among different liquids:

$$\tau = \mu \times \text{rate of shear strain}$$

In this equation, τ is the shearing resistance between the layers, and the rate of shear strain corresponds to the relative speed at which layer 1 slides by layer 2.

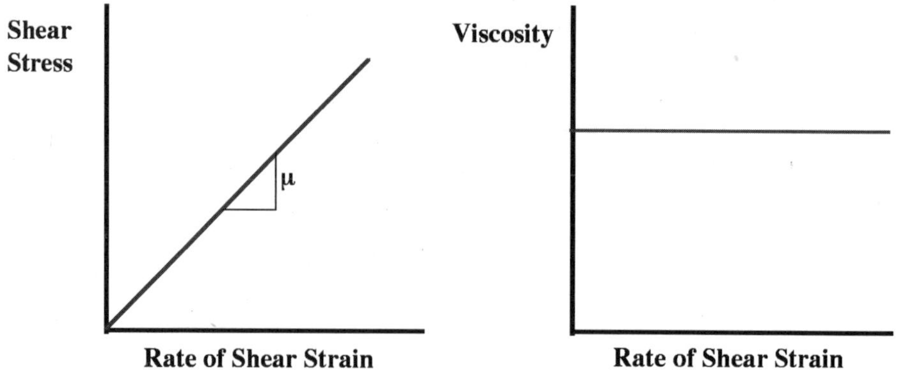

Figure 1.6 Newtonian Fluid Behavior

Asphalt binders typically display either Newtonian or non-Newtonian behavior as shear rate is changed. Newtonian fluids have a linear relationship between resisting force and relative velocity. In other words, if you apply twice the force to a Newtonian fluid, it will move twice as fast. Air, water, and hot asphalt (at temperatures greater than 60°C) are common Newtonian fluids. For these materials, viscosity (μ) is constant regardless of shear rate as shown in Figure 1.6.

Asphalt binders, particularly modified asphalts, may exhibit non-Newtonian behavior. Some asphalt materials may exhibit psuedoplastic or "shear thinning" behavior. This type of behavior, demonstrated in paints, is

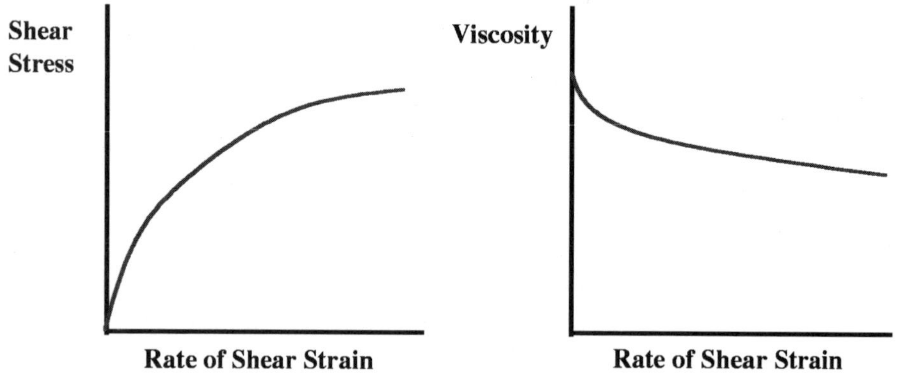

Figure 1.7 Pseudoplastic (Shear Thinning) Flow Behavior

characterized by a decrease in viscosity as shear rate is increased (Figure 1.7). In other words, the faster you stir the material, the thinner (more fluid) it gets. At moderate temperatures, asphalt exhibits "shear thinning" behaviors since its viscosity decreases at increasing shear rates.

Another type of non-Newtonian behavior, although less common, is dilatant or "shear thickening" behavior. This type of behavior, demonstrated in clay slurries, is characterized by an increase in viscosity as shear rate is increased (Figure 1.8). In other words, the faster you stir the material, the thicker (less fluid) it gets.

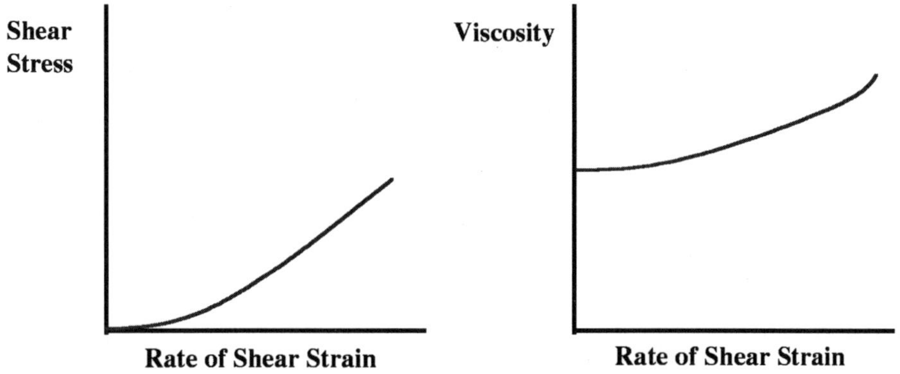

Figure 1.8 Dilatant (Shear Thickening) Flow Behavior

➤➤ *Low Temperature Behavior*

In cold climates (e.g., winter days) or under rapidly-applied loads (e.g., fast moving trucks), asphalt cement behaves like an elastic solid. Elastic solids are like rubber bands; when loaded they deform, and when unloaded, they return to their original shape.

If stressed beyond material capacity or strength, elastic solids may break. Even though asphalt cement is an elastic solid at low temperatures, it may become too brittle and crack when excessively loaded. For this reason, low temperature cracking sometimes occurs in asphalt pavement during cold weather. In these cases, loads are applied by internal stresses that accumulate in the asphalt pavement when it tries to shrink while being restrained by the lower pavement layers (e.g., when temperatures fall during and after a sudden cold front).

➤➤ *Intermediate Temperature Behavior*

Most environmental conditions lie between the extreme hot and cold situations. In these climates, asphalt binders exhibit the characteristics of both viscous liquids and elastic solids. Because of this range of behavior, asphalt is an excellent adhesive material to use in paving, but an extremely complicated material to understand and explain. When heated, asphalt acts as a lubricant, allowing the aggregate to be mixed, coated, and tightly-compacted to form a smooth, dense surface. After cooling, the asphalt acts as the glue to hold the aggregate together in a solid matrix. In this finished state, the behavior of the asphalt is termed viscoelastic; it has both elastic and viscous characteristics, depending on the temperature and rate of loading. Conceptually, this kind of response to load can be shown using the "spring-dashpot" model shown in Figure 1.9. Any force exerted on the asphalt causes a parallel reaction in both the spring and the dashpot. In hot mix asphalt, the spring represents the immediate elastic response of both the asphalt and the aggregate. The dashpot symbolizes the slower, viscous reaction of the asphalt, particularly in warmer temperatures. Most of the response is elastic or viscoelastic, (recoverable with time), while some of the response is plastic and non-recoverable.

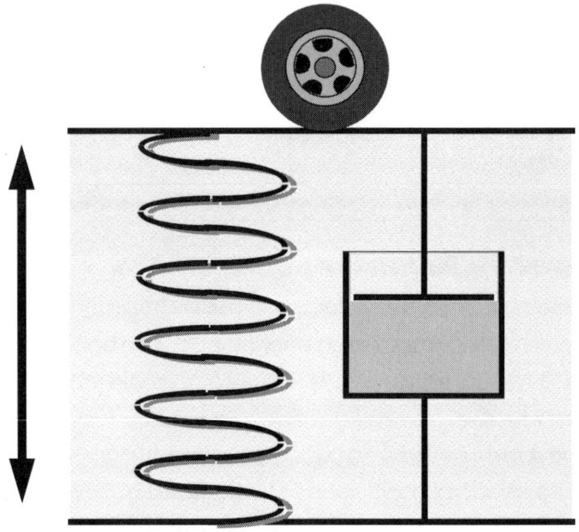

Figure 1.9 "Spring-Dashpot" Model of Viscoelastic Behavior

➤➤ *Aging Behavior*

Because asphalt cements are composed of organic molecules, they react with oxygen from the environment. This reaction is called oxidation and it changes the structure and composition of asphalt molecules. Oxidation causes the asphalt cement to become more brittle, generating the term oxidative hardening or age hardening. Oxidative hardening happens at a relatively slow rate in a pavement, although it occurs faster in warmer climates and during warmer seasons. Because of this hardening, old asphalt pavements are more susceptible to cracking. Improperly compacted asphalt pavements may exhibit premature oxidative hardening. In this case, inadequate compaction leaves a higher percentage of interconnected air voids, which allows more air to penetrate into the asphalt mixture, leading to more oxidative hardening.

In practice, a considerable amount of oxidative hardening occurs before the asphalt is placed. At the hot mixing facility, asphalt cement is added to the hot aggregate and the mixture is maintained at elevated temperatures for a period of time. Because the asphalt cement exists in thin films covering the aggregate, the oxidation reaction occurs at a much faster rate.

Other forms of hardening include volatilization and physical hardening. Volatilization occurs during hot mixing and construction, when volatile components tend to evaporate from the asphalt. Physical hardening occurs when asphalt cements have been exposed to low temperatures for long periods. When the temperature stabilizes at a constant low value, the asphalt cement continues to shrink and harden. Physical hardening is more pronounced at temperatures less than 0°C and must be considered when testing asphalt cements at very low temperatures.

CHAPTER 2

Asphalt Property Measurements

Because of its chemical complexities, asphalt specifications have been developed around physical property tests such as penetration, viscosity, and ductility. These physical property tests are performed at standard test temperatures, and the test results are used to determine if the material meets the specification criteria. However, there are limitations in what the results of these test procedures provide. Many of these tests are empirical, meaning that pavement performance experience is required before the test results yield meaningful information. Penetration is an example of this. The penetration test indicates the stiffness of the asphalt, but any relationship between asphalt penetration and performance has to be gained by experience. An additional drawback of empiricism is that the relationship between the test result and performance may not be very good.

Another limitation to these tests and specifications is that the tests do not give information for the entire range of typical pavement temperatures. Although viscosity is a fundamental measure of flow, it only provides information about higher temperature viscous behavior—the standard test temperatures are 60°C and 135°C. As well, penetration describes only the consistency at a medium temperature (25°C). Lower temperature elastic behavior cannot be realistically determined from this data to predict low temperature performance.

Penetration and viscosity specifications can classify different asphalts with the same grading, when in fact these asphalts may have very different temperature and performance characteristics. As an example, Figure 2.1 shows three asphalts that have the same viscosity grade because they are within the specified viscosity limits at 60°C, have the minimum penetration at 25°C, and reach the minimum viscosity at 135°C. While Asphalts A and B display the same temperature dependency, they have much different consistency at all temperatures. Asphalts A and C have the same consistency at low temperatures, but remarkably different high temperature consistency. Asphalt B has the same consistency at 60°C, but shares no other similarities with Asphalt C. Because these asphalts meet the same grade specifications, one might erroneously expect the same characteristics during construction and the same performance during hot and cold weather conditions.

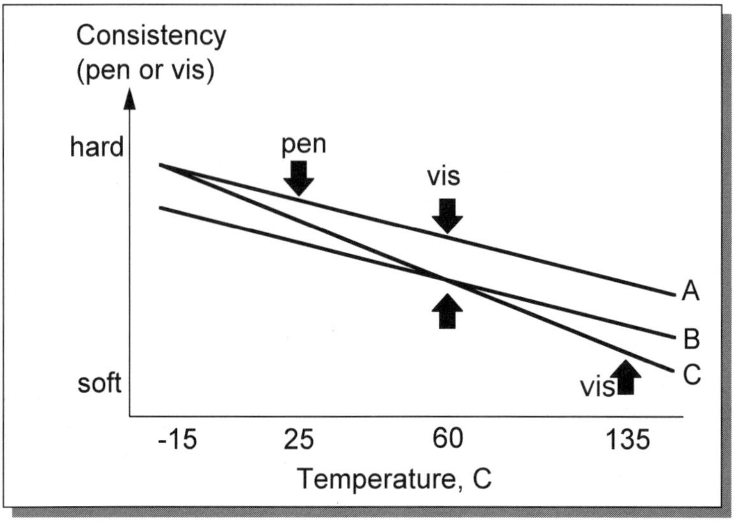

Figure 2.1 Variations of Three Viscosity-Graded Asphalts

Superpave Asphalt Binder Specification

Recognizing the deficiencies in the existing system, state highway agencies decided to fund a research program to develop a new system for specifying paving asphalt. In 1987, the Strategic Highway Research Program (SHRP) began developing new tests for measuring the physical properties of asphalt. A major result of this $50 million research effort was the Superpave binder specification. It is called a "binder" specification because it is intended for modified as well as unmodified asphalts. A unique feature of the Superpave specification is that the specified criteria remain constant, but the temperature at which the criteria must be achieved changes for the various grades. As an example, consider two construction projects—one at the equator and one at the Arctic Circle. Good asphalt performance is expected in both locations, but the temperature conditions under which specified binder properties must be achieved are vastly different.

The Superpave binder specification and the test methods used to characterize asphalt are being evaluated by both AASHTO and ASTM. AASHTO M320, ***Standard Specification for Performance-Graded Asphalt Binder***, standardizes the specification requirements. AASHTO's provisional standard, MP-1a, incorporates an alternate procedure for determining the critical low cracking temperature of an asphalt binder using a combination of two test procedures. Some researchers believe that the alternate approach in MP-1a is more rigorous and inclusive of most modified asphalts. These specifications may be revised as additional experience is gained.

The Superpave tests measure physical properties that can be related directly to field performance by engineering principles. The Superpave binder tests are also conducted at temperatures that are encountered by in-service pavements. Table 2.1 lists the binder test equipment and a brief description of why each test is used in the Superpave specification. Figure 2.2 describes how each test provides some indication of binder performance; however, the pavement structure and mixture properties will have some additional bearing on this performance.

Table 2.1 Superpave Binder Test Equipment

Equipment	Purpose
Rolling Thin Film Oven (RTFO) Pressure Aging Vessel (PAV)	Simulate binder aging (hardening) characteristics
Dynamic Shear Rheometer (DSR)	Measure binder stiffness and elasticity properties at high and intermediate temperatures (G^*, δ)
Rotational Viscometer (RV)	Measure binder viscosity at high temperatures
Bending Beam Rheometer (BBR) Direct Tension Tester (DTT)	Measure low temperature stiffness and failure properties

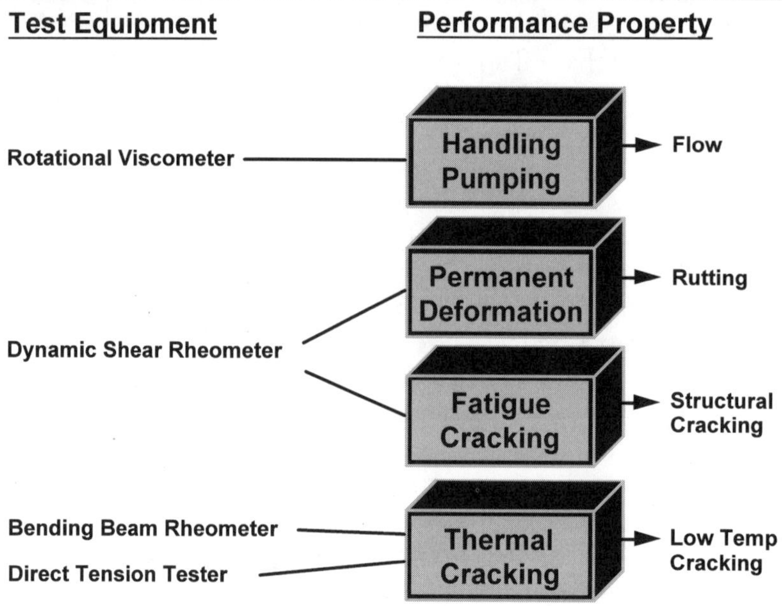

Figure 2.2 Superpave Laboratory Tests - Relation to Performance

The central theme of the Superpave binder specification is its reliance on testing asphalt binders in conditions that simulate the three critical stages during the binder's life. Tests performed on the original asphalt represent the first stage of transport, storage, and handling. The second stage represents the asphalt during mix production and construction and is simulated for the specification by aging the binder in a rolling thin film oven. This procedure exposes thin binder films to heat and air and approximates the aging of the asphalt during mixing and construction. The third stage occurs as the binder ages over a long period as part of the hot mix asphalt pavement layer. This stage is simulated for the specification by the pressure aging vessel. This procedure exposes binder samples to heat and pressure in order to simulate years of in-service aging in a pavement.

This chapter introduces the concepts and test procedures for the asphalt binder tests performed to determine compliance with the applicable specification. This information is intended to provide only an overview of the test procedures. Specific details are given in the appropriate AASHTO or ASTM test methods.

Superpave Binder Aging Procedures

As discussed in Chapter 1, asphalt binders age primarily due to two different mechanisms: volatilization of light oils present in the asphalt and oxidation by reacting with the oxygen in the environment. The blending and agitation in the hot mixing facility and during placement ages the binder by both mechanisms because of the high temperature and air flow involved in the process. The Superpave specification uses the rolling thin film oven (RTFO) procedure to simulate this form of aging. After the asphalt pavement is constructed, aging continues, but the oxidation mechanism dominates because of the relatively moderate temperatures of the environment. The Superpave specification uses the pressure aging vessel (PAV) procedure to simulate this in-service aging. It should be noted that binder samples aged in the PAV have already been aged in the RTFO. Consequently, PAV residue represents binder that has been exposed to all of the environmental conditions to which binders are subjected during production and service.

➤➤ *Rolling Thin Film Oven Procedure*

The RTFO procedure serves two purposes. The first is to provide an aged asphalt binder that can be used for further testing of physical properties. The second is to determine the mass quantity of volatiles lost from the asphalt during the process. Volatile mass loss is an indication of the aging that may occur in the asphalt during mixing and construction operations. Some asphalts even gain weight during the RTFO procedure because of the oxidative products formed. The RTFO continually exposes fresh films of binder to heat and air flow. In some cases, its rolling action keeps modifiers (e.g., some polymers) dispersed in the asphalt.

The RTFO procedure requires an electrically-heated convection oven (Figure 3.1). The oven contains a circular, vertical carriage that holds the sample bottles and rotates about its center. An air jet blows air into each sample bottle at its lowest position as it circulates in the carriage. The RTFO oven must be preheated to the aging temperature, 163°C, for a minimum of 16 hours prior to use. The thermostat should be set so that the oven will return to this temperature within 10 minutes after the sample bottles are placed and the door closed.

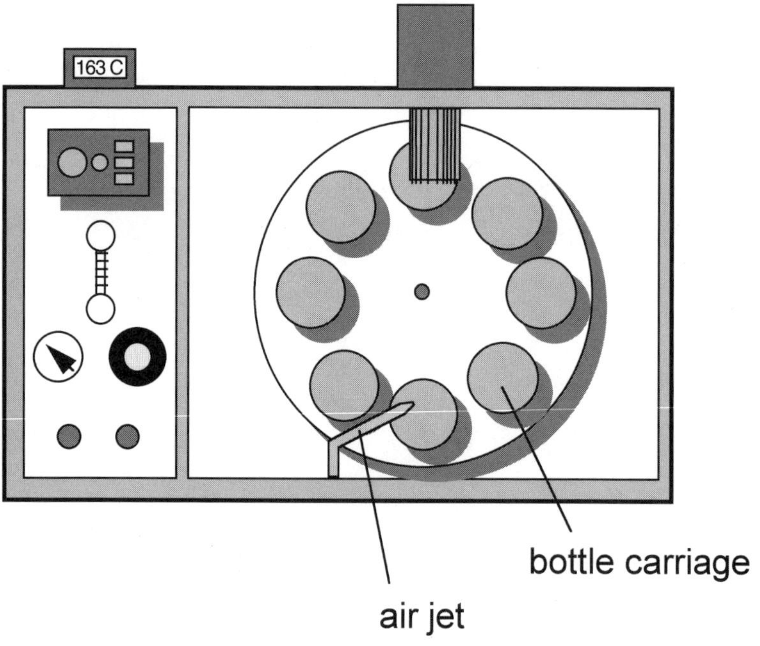

Figure 3.1 Rolling Thin Film Oven

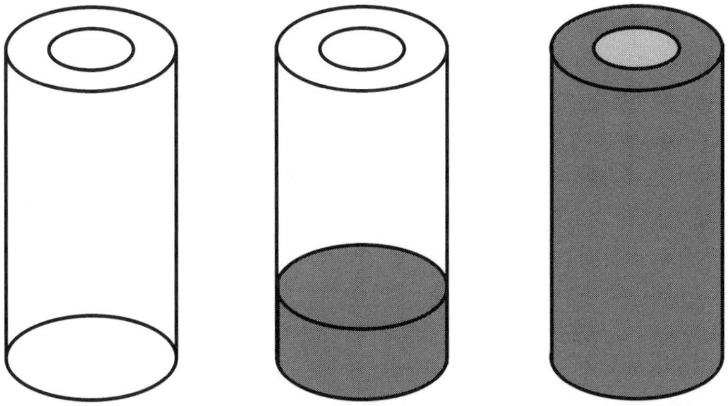

Figure 3.2 RTFO Bottle

To prepare for RTFO aging, a binder sample is heated until fluid, not exceeding 163°C. RTFO bottles are loaded with 35 grams of binder (Figure 3.2). Immediately after pouring the sample into the RTFO bottle, the bottle is turned on its side to a horizonal position and placed in a cooling rack. The bottles are allowed to cool 60 to 180 minutes. Eight sample bottles are required for Superpave binder testing. Two bottles are required to make the mass loss determination, and six bottles are used for further testing. These two mass-loss sample bottles are weighed to the nearest 0.001 gram.

Sample bottles are placed in the carriage and rotated at a rate of 15 revolutions per minute. The air flow is set at a rate of 4000 ml/min., and the samples are subjected to these conditions for 85 minutes.

After aging, the two bottles containing the mass loss samples are cooled, weighed to the nearest 0.001 gram, and the samples discarded. The RTFO residue from the other bottles is poured into a single container and stirred to achieve homogeneity. Scraping the coated bottles is necessary to ensure that enough material can be removed and a homogeneous sample is obtained. The RTFO-aged binder is then used for DSR testing, transferred into PAV pans for additional aging, or equally proportioned into small containers and stored for future use.

Mass change is reported as the average percent loss of the two samples after RTFO aging. It is calculated using this equation:

$$\text{Mass Change, \%} = \frac{\text{Aged mass - Original mass}}{\text{Original Mass}} \times 100$$

➤➤ *Pressure Aging Vessel*

As described previously, the effects of long term in-service aging of asphalt were not incorporated in specifications for asphalt binders prior to the Superpave specifications. The PAV, used for many years in asphalt research, was modified by SHRP and a new procedure was developed to simulate in-service aging. The PAV exposes the binder to high pressure and temperature for 20 hours to simulate the effects of long term aging. Since asphalt exposed to long term aging has also been through the mixing and construction process, the PAV procedure uses binder aged in the RTFO.

The pressure aging apparatus consists of the pressure aging vessel and a forced draft oven. Air pressure is provided by a cylinder of clean, dry compressed air with a pressure regulator, release valve, and a slow-release bleed valve. The pressure vessel is designed to operate under the pressure and temperature conditions of the test (2070 kPa and either 90°, 100°, or 110°C). The vessel must accommodate at least 10 sample pans and uses a sample rack that fits conveniently into the vessel. A continuous temperature monitor is required during the aging period. The oven should be able to control the internal temperature of the PAV to within ± 0.5°C during the aging period. A digital proportional control and readout of oven temperature is required.

Figure 3.3
Pressure
Aging Vessel

To prepare for the PAV, RTFO-aged binder is heated until fluid and stirred to ensure homogeneity. Three PAV sample pans of 50 grams each are prepared and the PAV pans are placed in the sample rack.

The unpressurized PAV is preheated to the desired test temperature. After preheating, the sample rack with the samples is placed in the hot vessel and the lid is quickly secured to avoid excessive heat loss. The aging process is conducted at different temperatures depending on the design climate. When the vessel temperature enters the range from 20°C to 2°C of the required temperature, the pressure is applied and the timing for the aging period begins.

After 20 hours, the pressure is gradually released— 8 to 10 minutes are usually required. If the pressure is released too quickly, foaming may occur. The sample rack is then removed from the PAV and the sample pans are removed from the rack and placed in an oven at 163°C for 15 minutes. The sample is then transferred to a storage container and degassed in a 170°C vacuum oven for 30 minutes at a pressure of 15 kPa absolute.

A report for the PAV includes sample identification; aging test temperature to the nearest 0.5°C; maximum and minimum aging temperature recorded to the nearest 0.1°C; total time during aging that temperature was outside the specified range to the nearest 0.1 minute; and total aging time in hours and minutes.

Superpave Binder Tests

As discussed previously, the Superpave binder tests are used to quantify the asphalt's performance at three stages of its life: in its original state, after mixing and construction, and after in-service aging. The RTFO test is used to simulate the binder aging that occurs during mixing and construction, and the PAV procedure is used to simulate the in-service aging. Table 3.1 shows the binder's aging condition used in the Superpave binder tests.

Table 3.1 Superpave Binder Test Aging Condition

Superpave Binder Test	Binder Condition
Dynamic Shear Rheometer (DSR)	Original binder
	RTFO-aged binder
	PAV-aged binder
Rotational Viscometer (RV)	Original binder
Bending Beam Rheometer (BBR)	PAV-aged binder
Direct Tension Tester (DTT)	PAV-aged binder

The relationships among the Superpave binder test results, the binder aging condition, and asphalt performance are further discussed in Chapter 4.

➤➤ *Dynamic Shear Rheometer*

Since asphalt behavior depends on both loading time and temperature, the ideal test for asphalt binders should include both factors. Fortunately, testing equipment with this capability already exists and is generically known as dynamic rheometers, dynamic shear rheometers, or oscillatory shear rheometers. Rheometers are adaptable for use with asphalt, so both time and temperature effects can be evaluated. When used to test asphalt binders, the dynamic shear rheometer (DSR) measures the rheological properties (complex shear modulus and phase angle) at intermediate to high temperatures.

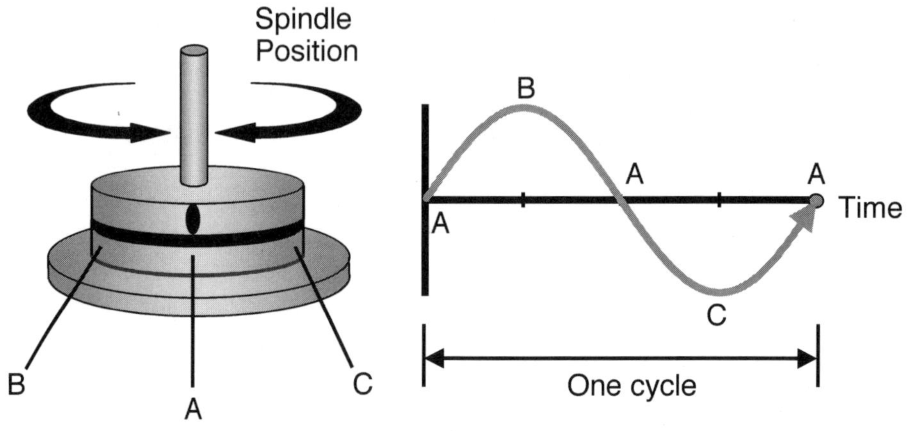

Figure 3.4 Dynamic Shear Rheometer Operation

The DSR operation is basic; asphalt is "sandwiched" between two parallel plates, one that is fixed and one that oscillates (Figure 3.4). As the plate oscillates, the centerline of the plate at point A (indicated by the dark vertical line) moves to point B. From point B, the plate centerline moves back and passes point A to point C. From point C the plate centerline moves back to point A. This oscillation is one cycle and is continuously repeated during the DSR operation. The speed of oscillation is frequency. All Superpave DSR binder tests are performed at a frequency of 10 radians per second, which is equal to approximately 1.59 Hz (cycles per second). During each cycle, both stress and strain are measured.

The DSR is used to characterize both viscous and elastic behavior by measuring the complex shear modulus (G*) and phase angle (δ) of asphalt binders. G* is a measure of the total resistance of a material to deformation when exposed to repeated pulses of shear stress. It consists of two components: elastic (recoverable) and viscous (non-recoverable). δ is an indicator of the relative amounts of recoverable and non-recoverable deformation. The value of G* ("G star") and δ ("delta") for asphalts are highly dependent on the temperature and frequency of loading. At high temperatures, asphalts behave like viscous fluids with no capacity for recovering or rebounding. In this case, the asphalt could be represented by the vertical axis (viscous component only) in Figure 3.5; there would be no elastic component of G*, since $\delta = 90°$. At very low temperatures, asphalts behave like elastic solids which rebound from deformation completely. This condition is represented by the horizontal axis (elastic component only) in Figure 3.5. In this case, there is no viscous component of G*, since $\delta = 0°$.

Under normal pavement temperatures and traffic loadings, asphalt binders act with the characteristics of both viscous liquids and elastic solids. By measuring G* and δ, the DSR provides a more complete picture of the behavior of asphalt at pavement service temperatures. The vector arrows in Figure 3.5, G^*_1 and G^*_2, represent the complex moduli of Asphalts 1 and 2. When these asphalts are loaded, part of their deformation is elastic (E) and part is viscous (V); therefore, asphalt is a viscoelastic material. Even though both asphalts in Figure 3.5 are viscoelastic and have the same G*, Asphalt 2 is more elastic than Asphalt 1, because of its smaller δ. Because Asphalt 2 has a larger elastic component, it will recover much more deformation from an applied load. This example clearly shows that G* alone cannot describe asphalt behavior. The value of δ is also needed.

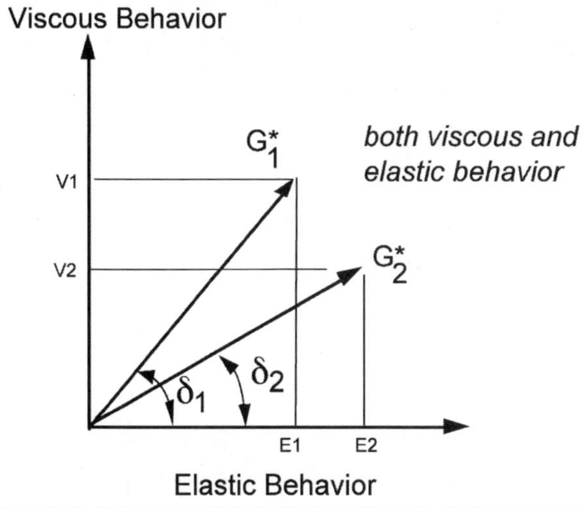

Figure 3.5
Viscoelastic Behavior

Specimen Preparation. A disk-shaped asphalt specimen with a diameter equal to the oscillating plate of the DSR is needed for testing. The thickness of the asphalt between the oscillating plate and the fixed plate must be carefully controlled. Proper specimen thickness is achieved by adjusting the gap between the two plates. This gap must be set before mounting the asphalt sample, but while the plates are mounted in the rheometer and at the test temperature. The gap is precisely adjusted by using a graduated micrometer wheel, delineated in microns.

Two different diameter oscillating plates, each with a corresponding gap thickness, are used in binder tests. The plate diameter and gap thickness used depend on the aged state of the asphalt being tested. Original (unaged) and RTFO-aged binders are tested with a 25-mm-diameter plate and a 1000 micron gap; PAV-aged binder is tested with an 8-mm-diameter plate and a 2000 micron gap. Before mounting the specimen, the gap is set at the desired value (1000 or 2000 microns) plus an extra 50 microns.

To prepare the test specimen, the asphalt binder is heated until fluid, stirring occasionally to remove air bubbles and achieve a homogeneous sample. Modified asphalt binders require higher temperatures than unmodified binders; but generally, the temperature should not exceed 163°C. There are different methods for preparing the specimen, but most commonly, the asphalt binder sample is poured into a silicon mold with the appropriate diameter and thickness for testing (see Figure 3.6.) Because lighter constituents of the asphalt binder sample may be absorbed by the silicone, care should be taken not to let the sample sit in the mold for any more than two hours before loading it in the DSR.

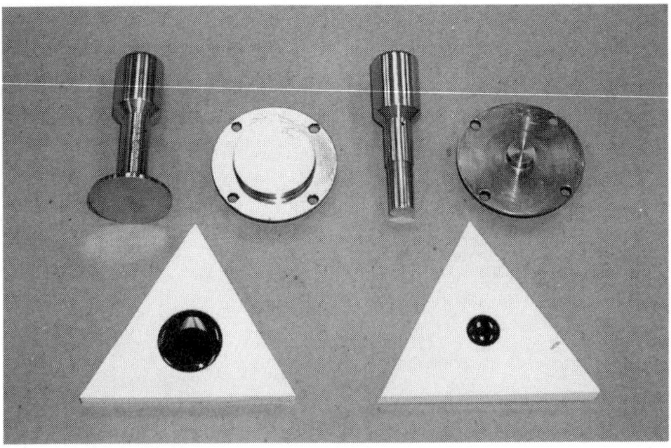

Figure 3.6 DSR Molds and Specimens, Plates and Spindles

After the asphalt is in place, the specimen is trimmed flush with the parallel plates, and the extra 50 microns is "dialed out" so that the gap is exactly at the desired value. The specimen will bulge slightly as shown in Figure 3.7.

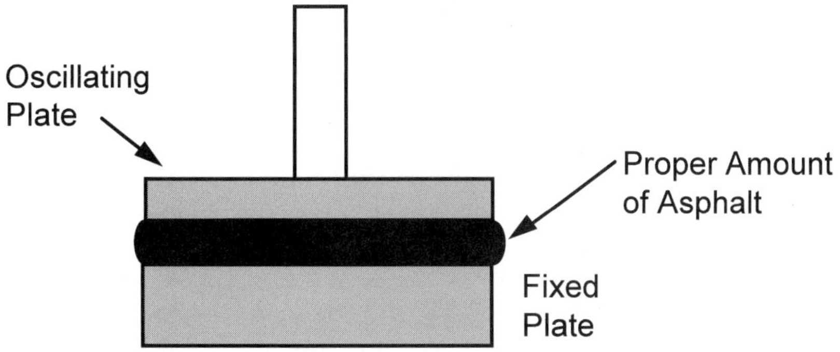

Figure 3.7 Asphalt Sample Configuration in DSR

Since asphalt binder properties are temperature dependent, rheometers must have a precise means of controlling the sample temperature. This is accomplished using either a circulating fluid bath or a forced air oven. Fluid baths typically use water to surround the sample. The water is circulated through a temperature controller that precisely adjusts and maintains the desired sample temperature. Air ovens surround the sample with air during testing. In either case, the temperature must be controlled so that the sample temperature is uniform and varies by no more than 0.1°C across the gap.

Test Procedure. After the asphalt specimen is in place and the test temperature stabilizes, a minimum of ten additional minutes is allowed for the specimen temperature to equilibrate. The actual temperature equilibration time is equipment dependent and should be checked using a calibration specimen with very accurate temperature sensing capabilities. A thermistor, wrapped with very thin silicone rubber sheeting material, is placed between the parallel plates to verify temperature.

A computer controls the DSR test parameters and records test results. Testing consists of setting the DSR to apply a constant oscillating stress and recording the resulting strain and time lag, δ. The Superpave specifications require that the oscillation speed is 10 radians/second. The operator sets the approximate value of shear strain ("strain amplitude"). Shear strain values vary from about one to 12 percent and depend on the aged state of the binder being tested. Original (unaged) binders and RTFO aged binders are tested at strain values of approximately ten to twelve percent. PAV-aged binders are tested at strain values of about one percent. In all cases, strain values must be small enough that the response of the binder (G*) remains in the linear viscoelastic range. In this range, G* is virtually unaffected by changes in strain level.

To begin the test, the sample is first conditioned by loading the specimen for 10 cycles. During this conditioning period, the rheometer measures the stress required to achieve the set shear strain and then precisely maintains this stress during the test. The shear strain can vary small amounts from the set value to achieve this constant stress. Variation in shear strain is controlled by the rheometer software. After the 10 conditioning cycles, ten additional cycles are applied to obtain test data. The rheometer software automatically computes and reports G* and δ, using the relationship between the applied stress and the resulting strain.

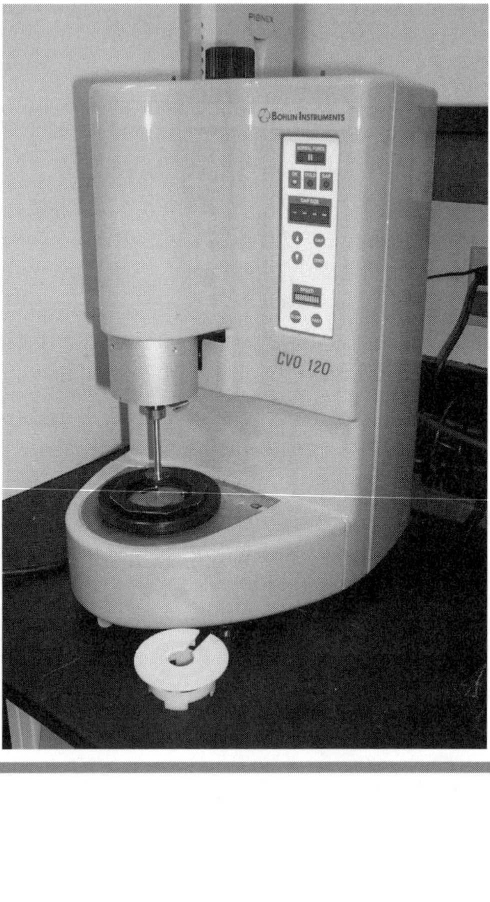

Figure 3.8
Dynamic Shear Rheometer

Data Analysis. Figure 3.9 shows the two extreme types of behavior, totally elastic and totally viscous.

The complex shear modulus, G*, is the ratio of total shear stress (τ_{max} - τ_{min}) to total shear strain (γ_{max} - γ_{min}). The time lag between the applied stress and the resulting strain (for constant stress rheometers as shown in Figure 3.7) or the applied strain and resulting stress (constant strain rheometers) is related to the phase angle, δ. For a perfectly elastic material, an applied load causes an immediate response; thus, the time lag or phase angle is zero. A viscous material (such as hot asphalt at mixing temperatures) has a relatively large time lag between load and response; in this case, the angle approaches 90 degrees. Asphalt binders are viscoelastic at normal pavement temperatures; they behave somewhere between the two extremes and the DSR displays a response resembling that shown in Figure 3.10.

The formulas used by the rheometer software to calculate τ_{max} and γ_{max} are shown in Figure 3.11.

Although the DSR is capable of providing much more information for analysis, only G* and δ are required for Superpave specifications. Two forms of G* and δ are used in the binder specification. Permanent deformation is governed by limiting G*/sin δ at the test temperatures to values greater than

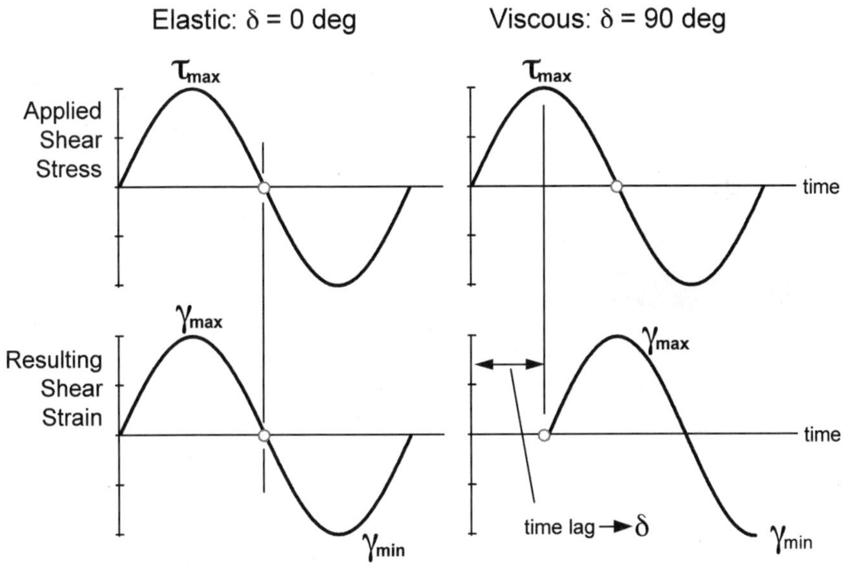

Figure 3.9 Stress-Strain Output for a Constant Stress Rheometer

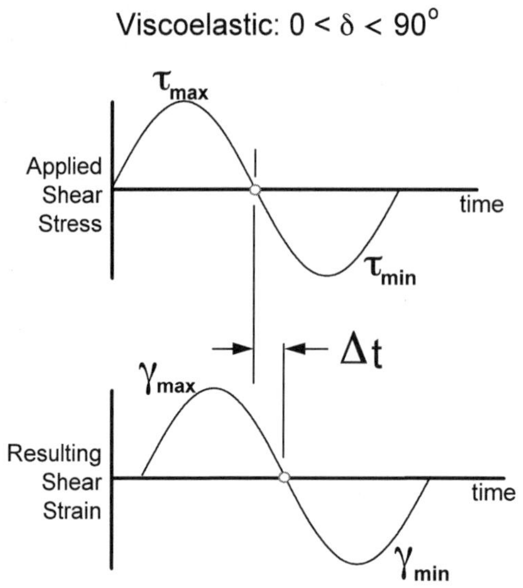

Figure 3.10 Stress-Strain Response of a Viscoelastic Material

$$G^* = \frac{\tau_{max} - \tau_{max}}{\gamma_{max} - \gamma_{max}}$$

Δt = time lag → δ

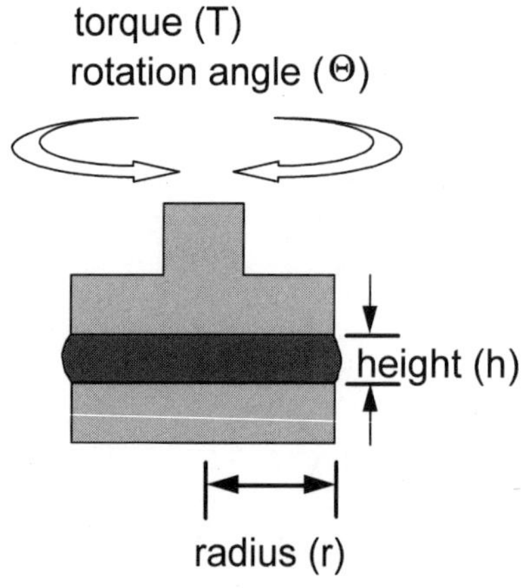

$$\tau = \frac{2T}{\pi r^3}$$

$$\gamma = \frac{\Theta r}{h}$$

Figure 3.11 DSR Asphalt Specimen Calculations

where T = maximum applied torque,
r = radius of specimen (either 12.5 or 4 mm),
θ = rotation angle, and
h = specimen height (either 1 or 2 mm).

Table 3.2 Example DSR Test Information

DSR Project - Superpave Binder Test	Strain Controlled
Parameters :	
Measurement Type :	Intermediate Temperature Range
Target Temperature :	22.0° C
Strain Amplitude :	1.00 percent
Plate Diameter :	8.0 mm
Plate Gap :	2.000 mm
Equilibration Time :	10.0 minutes
Ancillary Info :	
Operator ID :	PAT
Sample ID :	GRP3
Sample Type :	Pressure Aging Vessel Residue
Test Number :	0001
Measurement Results :	
Completed :	4/14/03 3:18 PM
Modulus (G*) :	3787 kPa
Phase Angle (delta) :	44.9 degrees
G* sin (delta) :	2671 kPa
Strain Amplitude :	1.00 percent
Final Temperature :	21.9° C
Osc. Frequency :	10.08 radians/second
Test Status :	PASSED

1.00 kPa for original binder and 2.20 kPa after RTFO aging. Fatigue cracking is governed by limiting G*sin δ of pressure aged material (in the PAV) to values less than 5000 kPa at the test temperature. More information describing how G* and δ are used in the Superpave binder specification is provided in Chapter 4.

Table 3.2 shows the format of test information for one DSR system.

➤➤ *Rotational Viscometer*

A rotational viscosity test is used to determine the flow characteristics of the asphalt binder to provide some assurance that it can be pumped and handled at the hot mixing facility. A rotational coaxial cylinder viscometer as shown in Figure 3.12 and described in AASHTO T316, *Viscosity Determination of Asphalt Binder Using Rotational Viscometer*, is necessary to evaluate the various types of asphalt binders. Unlike capillary tube viscometers, the

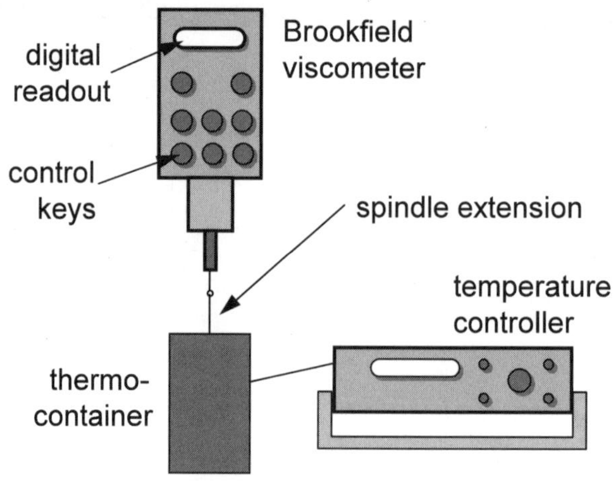

Figure 3.12 Rotational Viscometer

rotational viscometers have larger clearances between the components and, therefore, are applicable to modified and unmodified asphalts.

The rotational viscometer automatically calculates the viscosity at the test temperature. The rotational viscosity is determined by measuring the torque required to maintain a constant rotational speed of a cylindrical spindle while submerged in an asphalt binder sample at a constant temperature (Figure 3.13). This torque is directly related to the binder viscosity, which is calculated

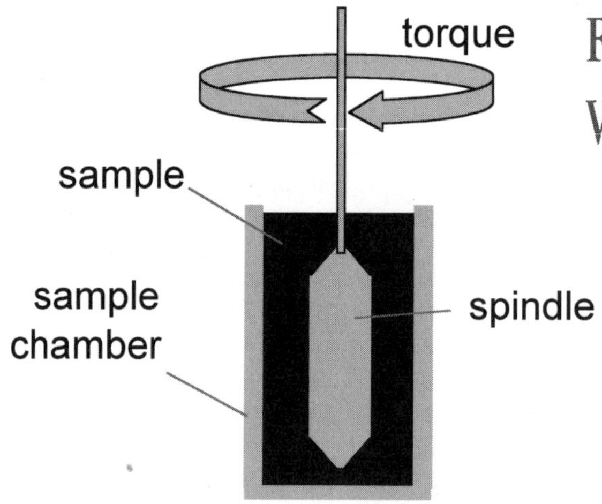

Figure 3.13 Rotational Viscometer Operation

automatically by the viscometer. Since this binder viscosity is used to ensure that the asphalt is fluid enough at normal operating temperatures to pump and mix with aggregate, it is measured on original or "tank" asphalt. The viscometer can also be used to develop temperature-viscosity charts for estimating mixing and compaction temperatures for use in mixture design.

Specimen Preparation. Approximately 30 grams of binder are heated in an oven until sufficiently fluid to pour. The sample should be stirred occasionally during heating to remove entrapped air. Asphalt is weighed into the preheated sample chamber. The amount of asphalt used is typically 8 to 11 grams and varies with the size of spindle. The sample chamber containing the binder sample is placed in the preheated thermo-container, the preheated spindle is lowered into the sample, and the binder is ready to test when the temperature stabilizes, usually within 30 minutes.

Test Procedure. The apparatus used to measure rotational viscosity consists of two parts: rotational viscometer and the temperature control system. The rotational viscometer consists of a motor, spindle, control keys, and digital

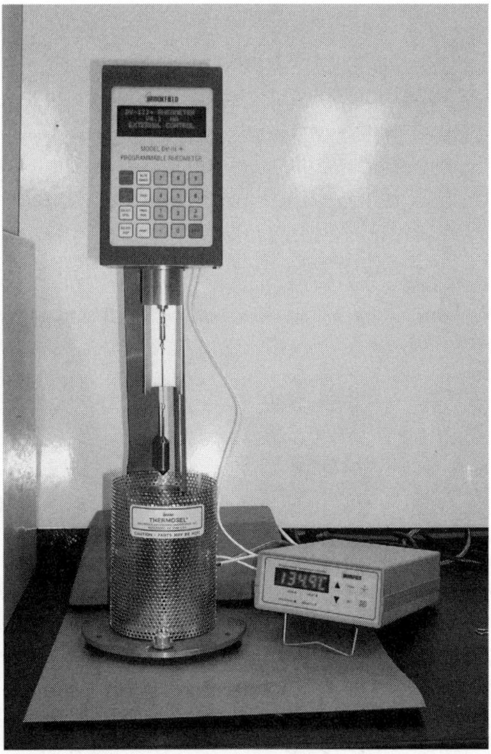

Figure 3.14 Rotational Viscometer

readout. The motor powers the spindle rotation through a torsional spring. The spring is wound as the torque increases. The torque in the spring is measured by a rotary transducer. For specification testing, the motor is set to operate at 20 rpm.

The spindle resembles a plumb bob and spindle rotation is resisted by the viscous binder. Many spindles are available for the rotational viscometer; the proper spindle is selected based on the viscosity of the binder being tested.

The temperature control system consists of the sample chamber, thermo-container, and temperature controller. The sample chamber is a stainless steel or aluminum cup; the thermo-container holds the sample chamber and consists of electric heating elements that are used to maintain or change the test temperature. The controller allows the test temperature to be set at the required 135°C.

To function properly, the viscometer and thermo-container must be leveled using bubble levels and leveling screws. Control keys are used to input test parameters such as spindle number, set rotational speed, and turn the motor on and off. The spindle is lowered into the chamber containing the hot sample and the spindle is coupled with the viscometer.

A waiting period of about 15 minutes may be needed to reach a uniform sample temperature of 135°C. During this period, the viscometer motor is turned on and the viscosity reading and the percent torque (should be between 2 and 98 percent) can be observed on the digital display. If the percent torque is outside of this range, a different size spindle will be required. As the temperature equalizes, the viscosity reading will stabilize and test results are recorded. The digital display is set to show the information that is needed for the report: viscosity, test temperature, spindle number, and speed. Three viscosity readings are recorded at 1-minute intervals. Figure 3.15 shows the four possible displays; only the upper left item in the display changes.

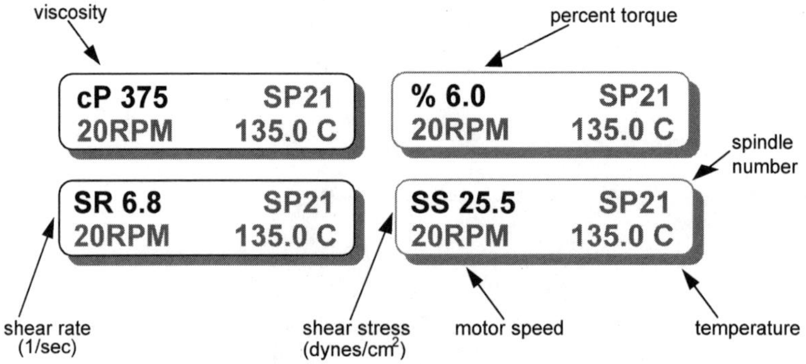

Figure 3.15 Rotational Viscometer Displays

It may be desirable to determine the binder viscosity at temperatures other than 135°C. For example, most agencies use equiviscous temperatures for mixing and compaction during mix design. Regardless of the grade used, the binder temperatures are adjusted to obtain the same specified range of binder viscosity when mixing with aggregate and compacting specimens in the laboratory. Therefore, a viscosity-temperature relationship for the binder is needed. To accomplish this, the temperature controller is reset to a higher desired temperature (e.g. 165°C), and the test is performed as before.

Data Analysis. The viscosity at 135°C is reported as the average of three readings. The digital output of some rotational viscometers is in units of centipoise (cP) while the Superpave binder specification uses Pascal-seconds, Pa•s. The conversion used is 1000 cP = 1 Pa•s. Therefore, to obtain the viscosity in Pa•s, the rotational viscosity in cP is multiplied by 0.001. As mentioned previously, the test temperature, spindle number, and speed are also required report items. The Superpave binder specification requirement of a maximum of 3 Pa•s is applied at the discretion of the specifying agency and may be waived if the binder supplier guarantees that the binder can be handled and pumped at the necessary temperatures.

➤➤ *Bending Beam Rheometer*

Asphalt binders at low temperatures are too stiff to reliably measure properties using the parallel plate geometry of the DSR. Therefore, SHRP researchers developed the bending beam rheometer (BBR) to accurately

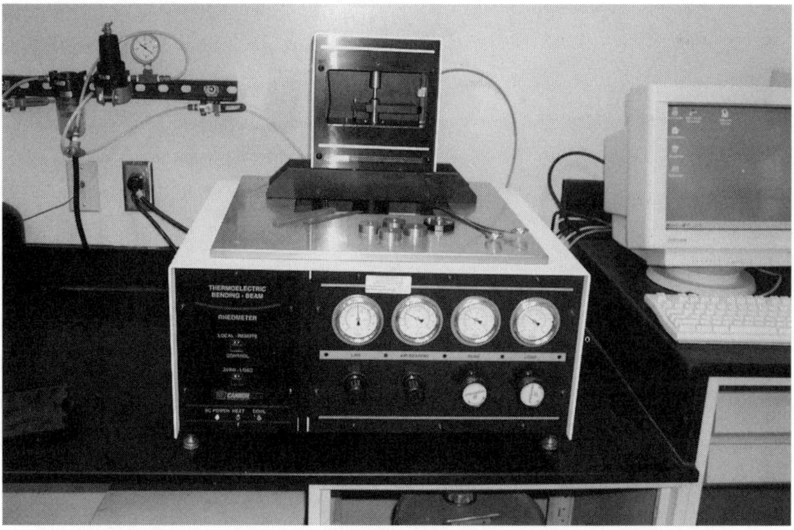

Figure 3.16 Bending Beam Rheometer

evaluate binder properties at low pavement temperatures. Used together, the dynamic shear and the bending beam tests provide stiffness behavior of asphalt binders over a wide range of temperatures. Although stiffness can also be used to estimate failure or strength properties, for some asphalt binders (especially modified asphalts,) the relationship between stiffness and strength properties is not well known. This is why some researchers believe that the alternate procedure for determining a critical low cracking temperture specified in AASHTO MP-1a represents a more rigorous approach than simply using BBR data. In the MP-1a alternate procedure, an additional test, the direct tension test, is conducted to measure strength and ability to stretch before breaking (strain at failure.) The direct tension test is covered in a later section.

The BBR is used to measure how much a binder deflects or creeps under a constant load at a constant temperature. The BBR test temperatures are related to a pavement's lowest service temperature, when the asphalt binder acts more like an elastic solid. Furthermore, the test is performed on binders that have been aged in both a rolling thin film oven and the pressure aging vessel. Therefore, the test measures the performance characteristics of binders as if they had been exposed to hot mixing in a mixing facility and some in-service aging.

The BBR gets its name from the test specimen geometry and loading method used during testing. The key elements of the BBR (Figure 3.16) are a loading frame, temperature-controlled fluid bath, and the computer control and data acquisition system. A blunt-nosed shaft applies load to the midpoint of the simply-supported asphalt beam. A load cell is mounted on the loading

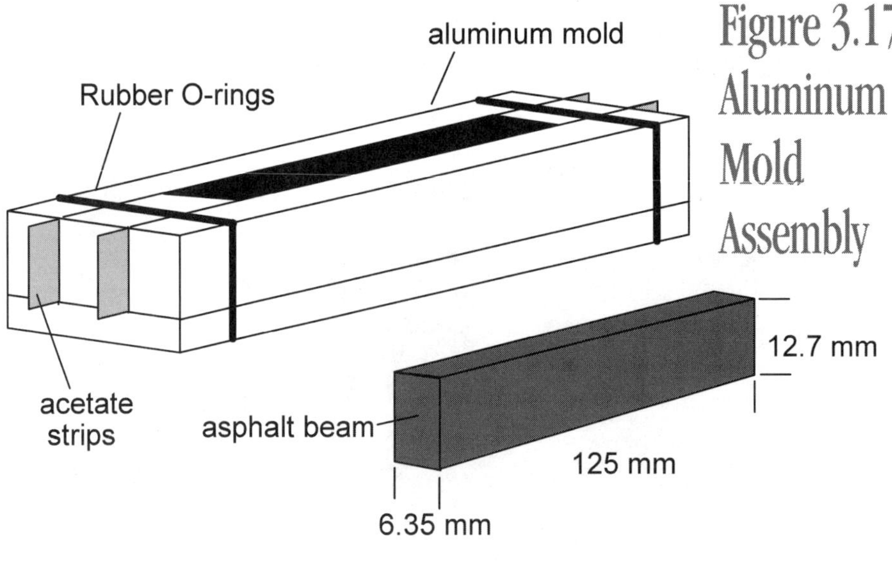

Figure 3.17 Aluminum Mold Assembly

shaft which is enclosed in an air bearing to eliminate friction during loading. A transducer on the loading shaft monitors deflection. Loads are applied by pneumatic pressure.

The test method uses beam theory to calculate the stiffness of an asphalt beam sample under a creep load. By applying a constant load to the asphalt beam and measuring the center deflection of the beam throughout the four-minute test procedure, the creep stiffness (s) and creep rate (m) can be calculated. The creep load simulates thermal stresses that gradually build up in a pavement when temperature drops. Creep stiffness is the resistance of the asphalt binder to creep loading and the m-value is the change in asphalt stiffness with time during loading.

The temperature bath contains a fluid that will not freeze at the test temperatures used. Fluids such as methanol, ethanol, ethylene glycol, water and combinations of these are used. The fluid must not freeze or be hard to circulate at normal test temperatures (down to -36°C). The fluid is circulated between the test bath and a circulating bath which controls the fluid temperature to within 0.1°C. Circulation or other bath agitation that may disturb the test specimen and influence the testing process must be avoided.

Specimen Preparation. Test specimens may be prepared using a rectangular aluminum mold (Figure 3.17). The mold assembly consists of aluminum base, side and end plates; acetate base and side strips; and rubber O-rings. Prior to assembling the mold, the inside surfaces of the two side plates and base plate are lightly greased with petroleum-based jelly. The three acetate strips are then placed against the greased faces. The end pieces are treated with a release agent composed of glycerin and talc. The mold is then assembled and held together with two rubber O-rings. The asphalt beam specimens are formed by heating the binder until fluid (usually about 135°C, but not to exceed 163°C) and pouring it into the mold from one end to the other in a continuous motion.

After a cooling period of about 45 to 60 minutes, excess asphalt is trimmed from the upper surface using a hot spatula. The asphalt specimen remains in the mold at room temperature, but not longer than two hours, and it is removed from the mold only when ready to begin the conditioning procedure. To demold the specimen, cool the specimen in a -5°C freezer or an ice bath for five to ten minutes. (The rheometer testing bath should not be used for cooling, since this may cause excessive fluctuations in the bath temperature.) After demolding, the beams must be handled carefully to avoid damage prior to testing.

The final preparation is temperature conditioning the beams in the BBR test bath at test temperature for 60 minutes. Tests on two beams are averaged for one test result. After temperature conditioning, the beams are

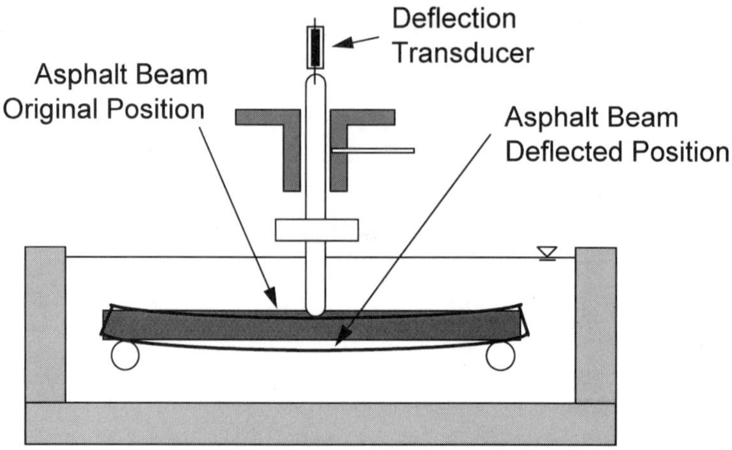

Figure 3.18 Bending Beam Test

tested. This testing tolerance requires the equipment and specimen preparation to be carefully coordinated.

Test Procedure. Hardware system calibration and compliance test procedures can be performed while the specimens are conditioned at test temperature in the testing bath. The deflection transducer and load cell are calibrated, and the compliance of the loading frame is checked with a rigid stainless steel reference beam. The temperature transducer is checked by using a calibrated thermometer. A thinner reference beam is also supplied that can be used to periodically check overall system performance. Most of the system calibration is controlled by the BBR software and instructions are easily followed from the computer screen. As part of this process, the desired levels for preload and loading are set using the thin reference beam.

After the 60-minute thermal conditioning period in the bath, the asphalt beam is placed on the supports. The beam is then subjected to a series of load-conditioning steps. A 35 mN (milliNewton) preload is manually applied to ensure that the beam is in firm contact with the supports. A 980-mN seating load is then automatically applied for one second by the BBR software. After this seating step, the load is reduced to the preload level for a 20-second recovery period.

After the 20-second recovery period, the actual test begins. A 980 mN load is applied to the beam for a total of 240 seconds. The deflection (Figure 3.18) is measured with the deflection transducer. During the test, graphs of load and deflection versus time are continuously generated on the computer screen for observation. After 240 seconds, the test load is automatically removed and the rheometer software calculates creep stiffness and creep rate.

Data Analysis. Beam theory is used to obtain creep stiffness of the asphalt in this test. The equation for calculating creep stiffness, S(t), is:

$$S(t) = \frac{PL^3}{4\,bh^3\delta(t)}$$

where,

S(t) = creep stiffness (MPa) at time, t,
P = applied constant load, N,
L = distance between beam supports, 102 mm,
b = beam width, 12.5 mm,
h = beam thickness, 6.25 mm, and
$\delta(t)$ = deflection (mm) at time, t.

The BBR software makes this calculation using deflection versus time. Figure 3.19 shows a sample graph and the procedure used to obtain data $\delta(t)$.

The desired value of creep stiffness is when the asphalt has been loaded for two hours at the minimum pavement design temperature. However, using the concept of time-temperature superposition, SHRP researchers confirmed that by raising the test temperature 10°C, an equal creep stiffness can be obtained after only a 60 second loading. The obvious benefit is that a test result can be measured in a much shorter testing time. Using the creep stiffness equation and deflection data (graphically illustrated in Figure 3.19), the BBR software calculates the stiffness at 60 seconds. AASHTO M320 requires that the creep stiffness at the specified grade temperature be less than or equal to 300 MPa at 60 seconds. However, if the creep stiffness is between 300 and 600 MPa and the m-value is greater than or equal to 0.300 (both parameters determined at 60 seconds,) then the direct tension test failure strain requirement may be used in lieu of the creep stiffness requirement.

The second parameter determined from the results of the bending beam test is the m-value. The m-value represents the rate of change in the creep stiffness, S(t), versus time. This value is also calculated automatically by the bending beam software. To obtain m-value, the stiffness is calculated at several loading times (Figure 3.19). The m-value is the slope of the log stiffness versus log time curve at any time, t. The Superpave specification requires that m-value be greater than or equal to 0.300 at 60 seconds.

Computer-generated output for the bending beam test includes all the required reporting items. The output format generally includes plots of deflection and load versus time, actual load and deflection values at various times, test parameters, and operator information. Table 3.3 shows typical information from one BBR software system and Figure 3.20 illustrates typical output plots.

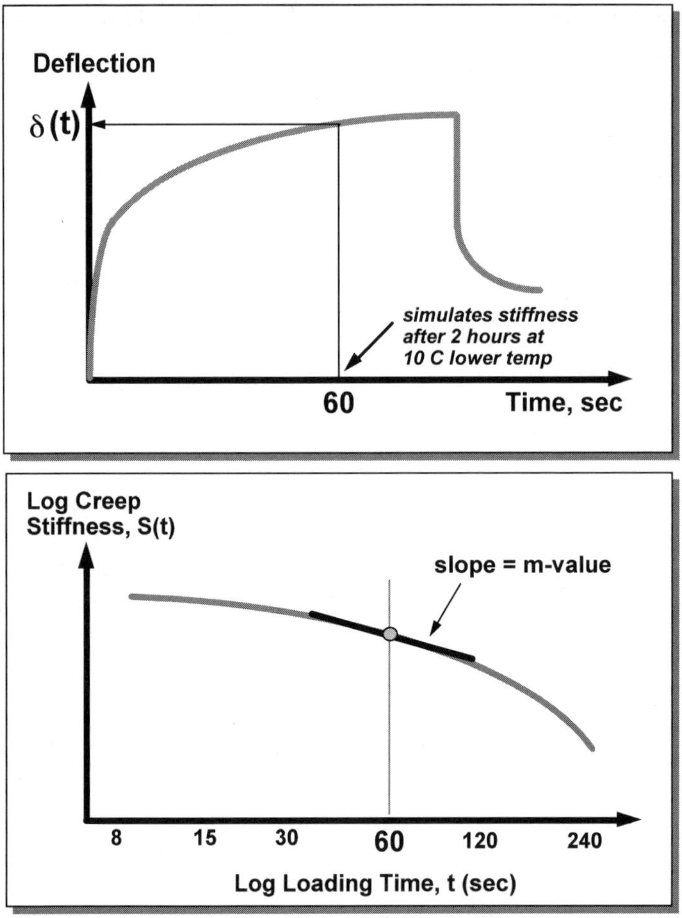

Figure 3.19 BBR Deflection and m-value

➤➤ *Direct Tension Tester*

Numerous past studies have shown that there is a strong relationship between the stiffness of conventional (unmodified) asphalt binders and the amount of stretching they undergo before breaking. Asphalts that undergo considerable stretching before failure are called "ductile"; those that break without much stretching are called "brittle". It is important that an asphalt binder be capable of a minimal amount of elongation. Typically, stiffer asphalts are more brittle and softer asphalts are more ductile.

Creep stiffness as measured by the BBR is not adequate to completely characterize the capacity of asphalts to stretch before breaking. For example, some binders exhibit high creep stiffness, but can also stretch farther before breaking. Consequently, SHRP researchers developed a specification system to accommodate these stiff-but-ductile binders. In the Superpave binder

Table 3.3 Example BBR Test Information

TEST INFORMATION

Proj: SPEX1	Target Temp: -18.0 C	Conf Test: 2.10e+08
Oper: PAT	Actual Temp: -18.0 C	Date: 04/05/03
Spec: IN-10	Soak time: 60.0 min	Load Const: 0.243112
Time: 15:53:47	Beam Width: 12.70 mm	Defl Const:.002417
Date: 04/05/03	Thickness: 6.35 mm	Date: 04/05/03
File: 0405037.DAT		

RESULTS

Time, sec	Force, N	Defl, mm	Measured Stiffness MPa	Estimated Stiffness MPa	Diff, %	m-value
8	.9658	.1877	414.8	414.3	-.1226	.261
15	.9678	.2231	349.8	349.7	-.00374	.277
30	.9684	.2730	286.0	286.8	.2903	.295
60	.9656	.3364	231.5	232.3	.3678	.313
120	.9665	.4152	187.7	185.9	-.9626	.331
240	.9662	.5327	146.2	146.9	.4417	.348

Regression Coefficients

a=5.829 b=-.2080 c=-.02949 R^2=.999821

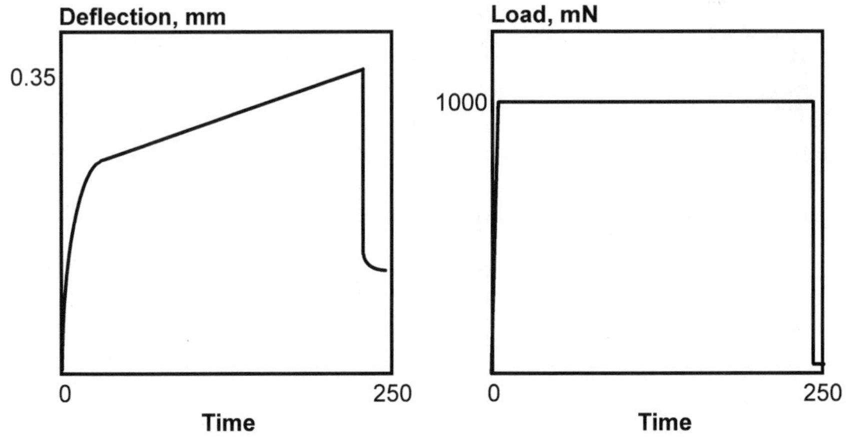

Figure 3.20 Typical BBR Deflection and Load Output

specification (AASHTO M320), these binders are allowed to have relatively high creep stiffness if they can also display reasonably ductile behavior at low temperatures. This additional requirement applies only to binders that have BBR creep stiffness between 300 and 600 MPa and have an m-value of 0.300 or greater. If the creep stiffness is below 300 MPa or the m-value is less than 0.300, the additional requirement and testing does not apply. While this particular system is used infrequently, it remains as an alternative to the more rigorous MP1a discussed in this section.

The equipment that measures the amount of binder stress and strain at failure at very low temperatures is the direct tension tester (DTT). The test is performed at the temperature range where asphalt generally exhibits brittle behavior (0C to -36C). Furthermore, the test is performed on binders that have been aged in both a rolling thin film oven and the pressure aging vessel. Consequently, the test measures the performance characteristics of binders as if they had been exposed to hot mixing in a mixing facility followed by some in-service aging.

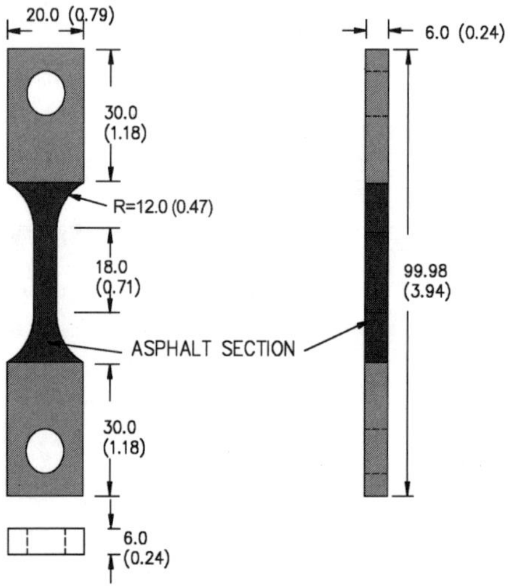

Figure 3.21 Superpave Direct Tension Specimen Geometry

In the direct tension test, a small "dog bone" shaped asphalt specimen (Figure 3.21) is pulled at a slow, constant rate until it fails. The elongation at failure is used to calculate the failure strain, which is an indication of whether the binder will behave in a brittle or ductile manner at the low test temperature (Figure 3.22).

The failure strain (ε_f) is the change in length (ΔL) divided by the effective gauge length (L_e). In the direct tension test, failure is defined as the load when the stress reaches its maximum value, not necessarily the load when

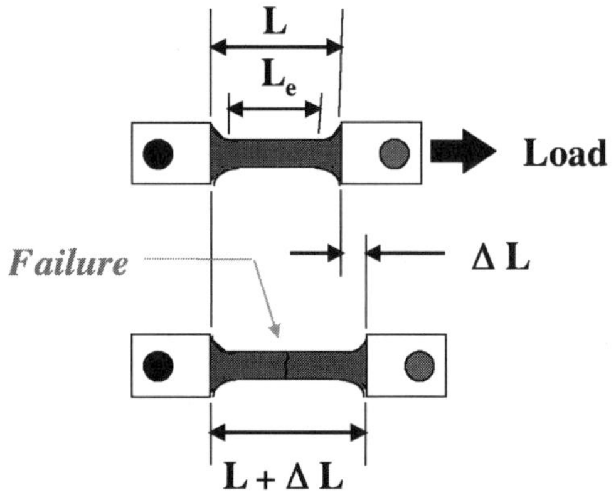

$$\text{failure strain } (\varepsilon_f) = \frac{\text{change in length } (\Delta L)}{\text{effective gauge length } L_e}$$

Figure 3.22
Determination of Failure Strain in the Direct Tension Test

the specimen breaks. Failure stress (σ_f) is the failure load divided by the original cross section of the specimen (36 mm^2). Figure 3.23 illustrates the stress-strain curve from a test conducted using the DTT.

Although stiffness has been used to estimate failure or strength properties, for some asphalt binders (especially modified asphalts) the relationship between stiffness and strength properties is not well known. This is why some researchers believe that the alternate procedure for determining a critical low cracking temperature specified in AASHTO MP-1a represents a more rigorous approach than AASHTO M320. In the MP-1a alternate procedure, the BBR data is used at all loading times, and with at least one temperature, to generate

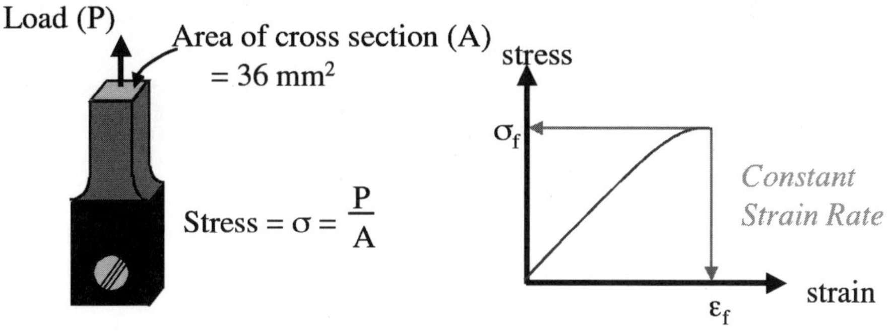

Figure 3.23
Determination of Failure Stress in the Direct Tension Test

a predicted thermal stress curve as a function of temperature. The DTT is used to determine the failure stress in an asphalt binder specimen when tensile fracture occurs. The failure tensile stress is plotted to determine whether, in the case of a verification test, a binder meets low temperature requirements or, in the case of a classification, the temperature at which the thermal stress curve from the BBR data exceeds the failure stress from the DTT. This intersection is termed the "critical cracking temperature" in the AASHTO MP-1a specification.

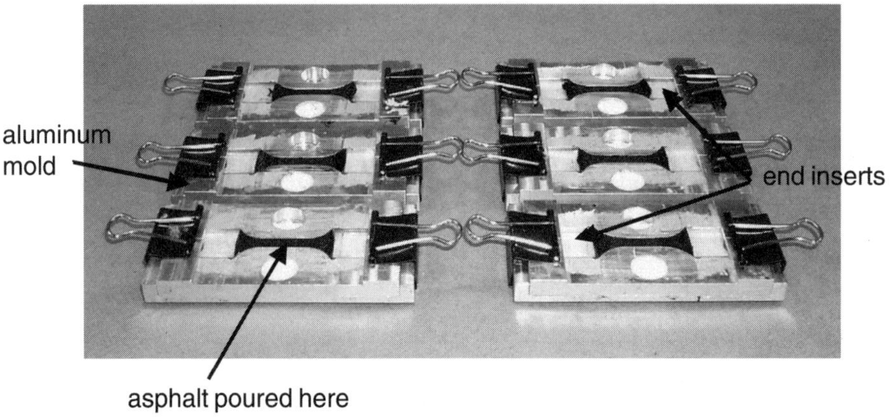

Figure 3.24 Direct Tension Specimens

Specimen Preparation. Direct tension specimens are formed in aluminum molds as shown in Figure 3.24. Six individual specimens are formed. One test result is produced from an average value of four specimens after discarding the two lowest stress-at-failure values. End inserts, made from phenolic G-10 and containing a 10-mm diameter hole inset with a stainless steel ring, are placed in the mold, the assembly and heat sink are preheated, and hot asphalt binder is then poured between the inserts in the space shown in Figure 3.24. Test specimens are poured and allowed to remain in the heat sink for a short period of time (5 minutes).

Test specimens have a mass of approximately 3 grams and are 100 mm long, including the end inserts. The dimensions of the plastic inserts and the formed binder test specimen are shown in Figure 3.21. The nominal binder cross section in the mid-length portion is 6 mm by 6 mm. A 12-mm radius is used to gradually widen the specimen to 20 mm at the end insert.

After the specimens are poured, they are allowed to cool at ambient temperature for 30-60 minutes before trimming (Figure 3.25). After trimming, the specimen is again allowed to cool for 10-15 minutes before demolding and

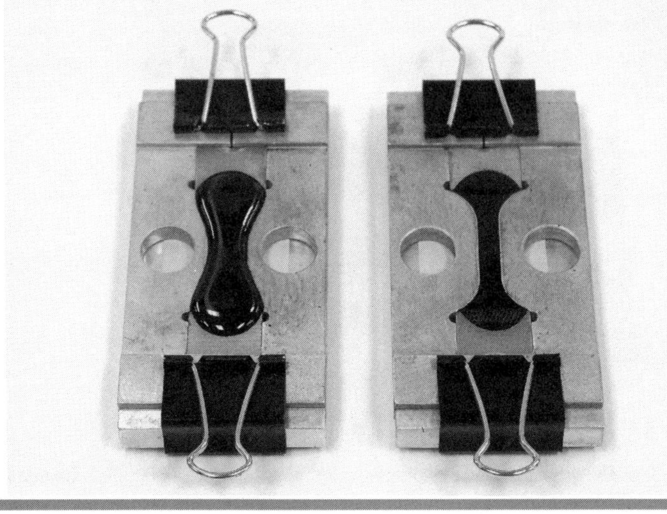

Figure 3.25
Direct Tension Test Specimen Before and After Trimming

conditioning in the fluid bath at the specified test temperature for 60 minutes prior to testing. This testing tolerance requires the equipment and specimen preparation to be carefully coordinated.

Test Procedure. Although the direct tension concept is simple, the equipment used requires complex features to accurately measure the very small strains involved in the test. The DTT consists of three components:

(1) an electromechanical test device to apply load,
(2) a transducer system to measure and control grip separation, and
(3) an environmental control system, usually in the form of a controlled temperature liquid bath.

The loading device must be capable of applying at least 500 N with a maximum crosshead travel of 20 millimeters at a rate of 1.0 mm/minute. Minimum and maximum deformation rates are required to be 0.001 mm/minute and 600 mm/minute, respectively. The measurement system for determining the elongation must have a resolution of 0.001 millimeter.

A key feature of the testing machine is the gripping system used to attach specimens to the alignment rods that apply the tensile load (Figure 3.26). The grips have a ball joint connection that ensures no bending within the binder test specimen.

The environmental control system includes an environmental chamber or liquid bath, and mechanical refrigeration unit capable of producing and precisely maintaining temperatures as low as -36°C. Most available liquid cooling units have been found to control adequately to only -36 — with the assistance of auxiliary stirring in the test chamber. In the case of an environmental chamber, air is used as the cooling medium. If a liquid bath is used, a solution of potassium acetate and water is used as the cooling medium.

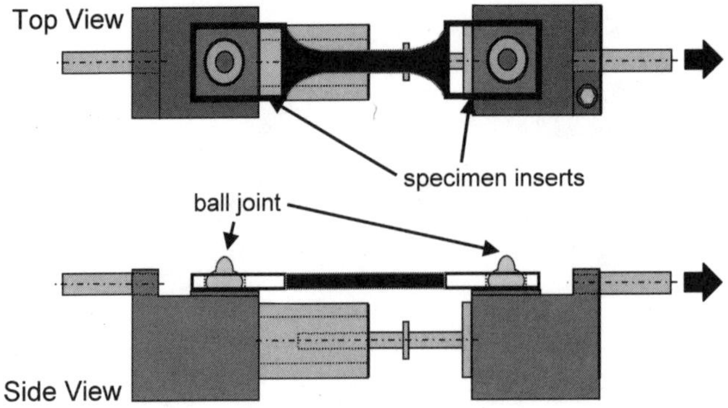

Figure 3.26 Direct Tension Test Loading System

At the end of the conditioning period (60 ± 5 minutes), each of the six specimens is tested individually. Before the test begins, the equipment is initialized and a small seating load is applied after mounting the specimen on the end grips. A tensile load is applied by pulling one end at a strain rate of 3% per minute (1.0 mm/minute) until the specimen fails. A test typically requires less than a minute from load application until failure. A test result is usually considered legitimate when fracture occurs within the 18 mm effective gauge section, which is the midsection of the specimen with the constant cross-section. A specimen that breaks near the end insert should be reported as an "end break".

Data Analysis. After dropping the two lowest stress-at-failure values, a single test result is generated by averaging the results from the four remaining test specimens. This practice is based on the concept that it is much more common in direct tension tests to have an early fracture because of specimen

Table 3.4 Typical DTT Test Results

	ε_f, %	σ_f, MPa	Max Load, N	Energy at Break, J	Max. Extension, mm	Start Temp, C	End Temp, C
1	1.05	4.74	173.90	0.03403	0.42	-17.9	-17.9
2	0.90	4.38	157.51	0.02667	0.37	-18.0	-18.1
3	0.95	4.67	168.24	0.03023	0.39	-18.1	-18.2
4	0.99	4.65	167.51	0.03142	0.40	-18.1	-18.2
5	0.90	4.24	155.15	0.02643	0.36	-18.1	-18.1
6	0.81	4.02	144.90	0.02198	0.33	-18.1	-18.1
Mean	0.93	4.45	161.20	0.02846	0.38	-18.0	-18.1

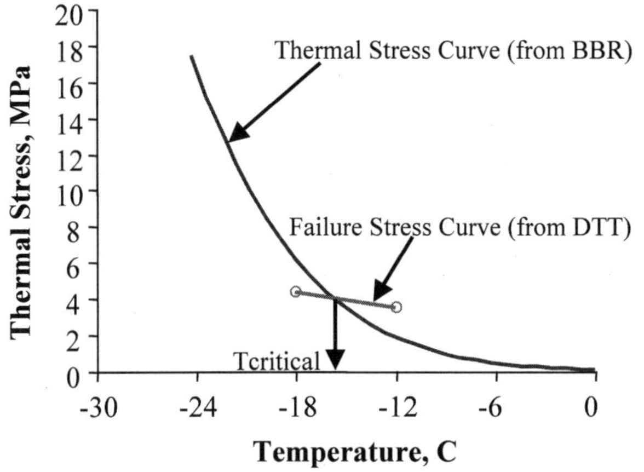

Figure 3.27 Thermal Stress Curve for Determining Critical Cracking Temperature

or procedural imperfections than an exceptionally "strong" test specimen that does not fracture early. Other analysis procedures may dictate different methods for evaluating the quality of the test data and its suitability for inclusion in the general average.

A single test result consists of the average strain to failure of multiple DTT specimens , reported to the nearest 0.01 percent and the average stress at failure reported to the nearest 0.01MPa. Table 3.4 shows typical DTT test output.

In Table 3.4, after dropping the two low values, the average strain at failure is 0.97% and the peak stress is 4.61 MPa. If the direct tension test is being performed for AASHTO M320, only the average strain at failure is needed to determine specification compliance. In AASHTO MP-1a, the peak stress is used as an input to calculate the critical cracking temperature of the asphalt binder. To determine critical cracking temperature, BBR test data is used to generate a thermal stress curve as a function of temperature. The DTT peak stress is then used to determine when tensile fracture occurs in an asphalt binder specimen. The peak tensile stress is plotted to determine at what temperature the thermal stress curve from the BBR data exceeds the tensile stress from the DTT. This intersection is termed the “critical cracking temperature” in the AASHTO MP-1a specification (Figure 3.27), and is used as an alternate method of determining the low temperature grade of an asphalt binder.

Although not used to determine specification compliance, these values are also required reporting items:

- Test temperature to the nearest 0.1C,
- Average rate of elongation to the nearest 0.01 mm/min,
- Peak load to the nearest N, and
- Type of break observed (brittle, brittle-ductile, or no break).

The Superpave asphalt binder specification is intended to improve performance by limiting the potential for the asphalt binder to contribute toward permanent deformation, low temperature cracking and fatigue cracking in asphalt pavements. The specification provides for this improvement by designating various physical properties with the equipment described in Chapter 3. This section briefly explains each of the new test parameters as they relate to pavement performance.

One important distinction between typical asphalt specifications and the Superpave specification is the overall format of the requirements. The required physical properties remain constant for all of the performance grades (PG). However, the temperatures at which these properties must be reached vary depending on the climate in which the binder is expected to be used. For example, the partial view of the Superpave specification format shown in Figure 4.1 shows that a PG 52-40 grade binder is designed to sustain the conditions of an environment where the average seven-day maximum pavement temperature is 52°C and the minimum pavement design temperature is -40°C.

The following sections describe each specified property, in the order in which they appear in the specification.

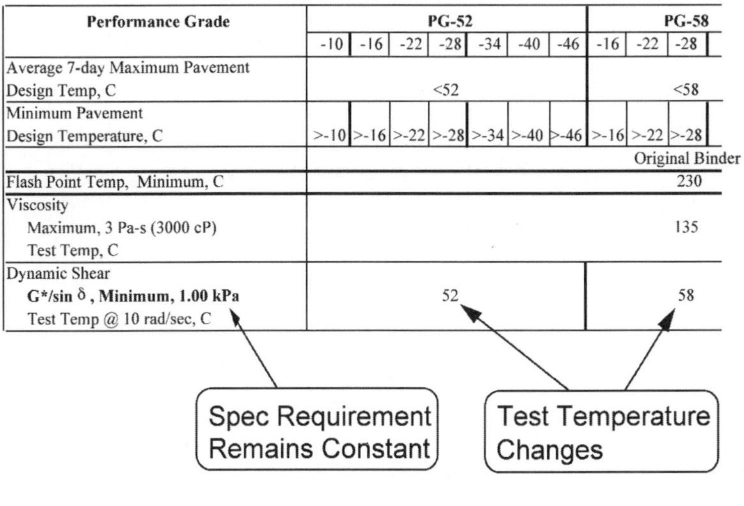

Performance Grade	PG-52							PG-58		
	-10	-16	-22	-28	-34	-40	-46	-16	-22	-28
Average 7-day Maximum Pavement Design Temp, C				<52						<58
Minimum Pavement Design Temperature, C	>-10	>-16	>-22	>-28	>-34	>-40	>-46	>-16	>-22	>-28
										Original Binder
Flash Point Temp, Minimum, C										230
Viscosity Maximum, 3 Pa-s (3000 cP) Test Temp, C										135
Dynamic Shear **G*/sin δ, Minimum, 1.00 kPa** Test Temp @ 10 rad/sec, C				52						58

Figure 4.1
Superpave Binder Specification Format

Safety

The flash point test, AASHTO T48, ***Standard Method of Test for Flash and Fire Points by Cleveland Open Cup***, indicates the temperature to which binder material may be heated without danger of instantaneous flash in the presence of an open flame. The flash point is well below the fire point, the temperature at which the asphalt will actually burn. (This fire point is seldom included in asphalt specifications.) The minimum temperature required for all grades is 230°C. This test is performed on the unaged binder.

Pumping and Handling

To ensure that asphalt binders, especially modified asphalts, can be pumped and handled at the hot mixing facility, the specification contains a maximum viscosity requirement on the unaged binder. This value is 3 Pa•s and it must be achieved at 135°C for all grades. AASHTO T316, *Viscosity Determination of Asphalts Binders Using Rotational Viscometer* is the specified test procedure. The specifying agency may waive this requirement if the binder supplier warrants that the binder can be adequately pumped and mixed at the specified temperatures.

Permanent Deformation

As discussed in the section describing the DSR in Chapter 3, the total response of asphalt binders to load consists of two components: elastic (recoverable) and viscous (non-recoverable). Pavement rutting or permanent deformation is the accumulation of the non-recoverable component of the responses to load repetitions at high service temperatures. The Superpave specification defines and places requirements on a rutting factor, $G^*/\sin\delta$, which represents a measure of the high temperature stiffness or rutting resistance of the asphalt binder. This factor is called "G star over sine delta." It is determined by dividing the complex shear modulus (G^*) by the sine of the phase angle (δ), which are both measured using the dynamic shear rheometer. To minimize rutting, $G^*/\sin\delta$ must be a minimum of 1.00 kPa for the original asphalt binder and 2.20 kPa after aging the binder using the RTFO procedure (Figure 4.2). High values of G^* and low values of δ are considered desirable attributes from the standpoint of rutting resistance. Thus, the Superpave specification promotes the use of stiff, elastic binders (unaged and RTFO aged) to address rutting resistance.

Excessive Aging

A mass loss requirement is specified to guard against using an asphalt binder that would age excessively from volatilization during hot mixing and construction. The mass loss requirement is calculated using the RTFO procedure—AASHTO T 240, *Standard Method of Test for Effect of Heat and Air*

Performance Grade
Average 7-day Maximum Pavement Design Temperature, C
Minimum Pavement Design Temperature, C
Flash Point Temp, minimum C
Viscosity Maximum, 3 Pa-s (3000 cP), Test Temp, C
Dynamic Shear **G*/sin δ, Minimum, 1.00 kPa** Test Temperature @ 10 rad/s, C
Rolling Thin Film Oven
Mass Loss, Maximum, %
Dynamic Shear **G*/sin δ, Minimum, 2.20 kPa** Test Temp @ 10 rad/sec, C

Spec Requirements to Address Rutting

Figure 4.2 Superpave Specification Rutting Factor Requirements

on a Moving Film of Asphalt (Rolling Thin Film Oven Test). The mass loss for any grade must not exceed 1.00 percent.

Fatigue Cracking

Like permanent deformation, G* and δ are also used in the Superpave binder specification to help control the fatigue of asphalt pavements. Since fatigue generally occurs at low to moderate pavement temperatures after the pavement has been in service for a period of time, the specification addresses these properties using binder aged in both the RTFO and PAV.

The DSR is again used to generate G* and δ. However, instead of dividing the two parameters, the two are multiplied to produce a factor related to fatigue. The fatigue cracking factor is G*sin δ, which is stated "G star sine delta." It is the product of the complex shear modulus, G*, and the sine of the phase angle, δ. The Superpave binder specification has a maximum value of 5000 kPa for G*sin δ (Figure 4.3). Low values of G* and δ are considered desirable attributes from the standpoint of resistance to fatigue cracking. Thus, the Superpave specification promotes the use of compliant, elastic binders (PAV aged) to address fatigue cracking.

Low Temperature Cracking

When the pavement temperature decreases, asphalt concrete shrinks. As the temperature drops, the asphalt binder contracts to a much greater degree than the aggregate in an asphalt concrete pavement, thus causing thermal stresses to develop in the pavement. When these stresses exceed the tensile strength of the asphalt mixture then a low temperature crack develops.

PAV Aging Temp, C
Dynamic Shear
G*sin δ, Maximum, 5000 kPa
Test Temp @ 10 rad/sec, C
Physical Hardening
Creep Stiffness
S, Maximum, 300 MPa
m-value, Minimum, 0.300
Test Temp, @60 sec, C
Direct Tension
Failure Strain, Minimum, 1.0%
Test Temp @ 1.0 mm/min, C

Specification requirement to address fatigue cracking

Figure 4.3 Superpave Specification Fatigue Cracking Factor Requirements

PAV Aging Temp, C
Dynamic Shear
G*sin δ, Maximum, 5000 kPa
Test Temp @ 10 rad/sec, C
Physical Hardening
Creep Stiffness
S, Maximum, 300 MPa
m-value, Minimum, 0.300
Test Temp, @60 sec, C
Direct Tension
Failure Strain, Minimum, 1.0%
Test Temp @ 1.0 mm/min, C

Specification requirements to address low temperature cracking

Figure 4.4 Superpave Specification Low Temperature Requirements

In AASHTO M320, the main way of examining the propensity of an asphalt binder to develop thermal stresses at a specified temperature is to use data generated from the bending beam rheometer (BBR). The BBR is used to apply a small creep load, at a specified temperature, to an asphalt binder sample molded into a beam shape and then measure the deformation under load as a function of time. By knowing the load and the beam dimensions, the creep stiffness (binder's resistance to load) can be calculated at any time. If the stiffness is too high, the asphalt binder sample will behave in a brittle manner, indicating that cracking is more likely to occur at that temperature. To prevent thermal cracking, creep stiffness has a maximum limit of 300 MPa (Figure 4.4). Since low temperature cracking usually occurs after the pavement has been in service for some time, this part of the specification addresses these low temperature properties using asphalt binder that has been aged in both the RTFO and PAV.

The rate at which binder stiffness changes with time at low temperatures is regulated through the m-value. In the AASHTO M320 specification, a high m-value is desirable, because as the temperature decreases and contraction occurs, the asphalt binder will respond as a material that is less stiff. This decrease in stiffness (ratio of stress over strain) leads to smaller tensile stresses in the asphalt binder and less chance for low temperature cracking. A minimum m-value of 0.300 after 60 seconds of loading is required by the Superpave asphalt binder specification (Figure 4.4).

Past studies have also indicated that if a binder can stretch at least one percent of its original length during this thermal contraction period, cracks are less likely to occur. As a result, the direct tension test (DTT) is included in AASHTO M320 as an alternative requirement to simply using creep stiffness and m-value. The DTT is used to pull an asphalt binder specimen in tension at a very slow rate, simulating the pavement condition as shrinkage occurs. The amount of strain that occurs before the specimen breaks is the failure, or peak, strain. To meet the requirements of AASHTO M320, this failure strain must be a minimum of 1.0 percent (Figure 4.4). In this specification, the DTT requirement only applies to asphalt binders that have a creep stiffness greater than 300 MPa and less than 600 MPa, with an m-value of 0.300 or greater. If the creep stiffness at the specified temperature is 300 MPa or less, then the direct tension test is not required.

Although stiffness can also be used to estimate failure or strength properties, for some asphalt binders (especially modified asphalts) the relationship between stiffness and strength properties is not well known. This is why some researchers believe that the alternate procedure for determining a critical low cracking temperature specified in AASHTO MP-1a represents a more rigorous approach than AASHTO M320. In the MP-1a alternate procedure, the BBR data is used at all loading times (and one or more temperatures) to generate a predicted thermal stress curve as a function of temperature. The DTT is then used to determine the failure stress in an asphalt binder specimen when tensile fracture occurs. The failure tensile stress is plotted to determine at what temperature the thermal stress curve from the BBR data exceeds the failure stress from the DTT. This intersection is termed the "critical cracking temperature" in the AASHTO MP-1a specification, and is used an alternate method of determining the low temperature grade of an asphalt binder. In the AASHTO MP-1a specification, an asphalt binder with a low temperature grade of -22C would need to have a critical cracking temperature of -22C or lower.

Contrary to the previous grading systems, the Superpave binder specification is theoretically based directly on performance rather than on empirical relationships between basic physical properties and observed performance. Performance graded binders are selected based on the climate in which the pavement will serve. Unlike all other systems, the physical property requirements are constant among all performance grades. The distinction among the various binder grades is the specified minimum and maximum temperatures at which the requirements must be met. For example, a binder classified as a PG 58-34 means that the binder will meet the high temperature physical property requirements up to a temperature of 58°C and the low temperature physical property requirements down to -34°C.

AASHTO M320 contains a listing of the more common PG grades. However, the PG grades are not limited to those given classifications. In actuality, the specification temperatures are unlimited, extending unbounded in both directions. The high and low temperatures extend as far as necessary in the standard six-degree increments. For example, even though a PG 58-10 is not shown, it exists as a legitimate grade in the system.

Even with binder grades classified according to high and low temperature categories, more information is needed to select a binder for a particular location. The LTPPbind Superpave software, developed by the Federal Highway Administration, assists users in selecting binder grades. The Superpave software contains three methods by which the user can select an asphalt binder grade:

Geographic Area: An agency would develop a map showing binder grade to be used based on weather and/or policy decisions.

Pavement Temperature: The designer determines design pavement temperature.

Air Temperature: The designer determines design air temperatures, which are converted to design pavement temperatures.

The Superpave software assists in the third method by providing a database of weather information for 6092 reporting weather stations in the US and Canada that allows users to select binder grades for the climate at the project location. For each year that these weather stations have been in operation, the hottest seven-day period was identified and the average maximum air temperature for this seven-day period was calculated. SHRP researchers selected this seven-day average value as the optimum method to characterize the high temperature design condition. For all the years of operation, the mean and standard deviation of the seven-day average maximum air temperature have been computed. Similarly, the one-day minimum air temperature of each year was identified and the mean and standard deviation of all the years of record was calculated. Weather stations with less than 20 years of operations were not used.

However, the design temperatures to be used for selecting asphalt binder grade are the pavement temperatures, not the air temperatures. For surface layers, Superpave defines the high pavement design temperature at a depth 20 mm below the pavement surface, and the low pavement design temperature at the pavement surface.

Using theoretical analyses of actual conditions performed with models for net heat flow and energy balance, and assuming typical values for solar absorption (0.90), radiation transmission through air (0.81), atmospheric radiation (0.70), and wind speed (4.5 m/sec), this equation was developed for the high pavement design temperature:

$$T_{20mm} = (T_{air} - 0.00618\,Lat^2 + 0.2289\,Lat + 42.2)(0.9545) - 17.78$$

where T_{20mm} = high pavement design temperature at a depth of 20 mm
T_{air} = seven-day average high air temperature, °C
Lat = the geographical latitude of the project in degrees.

The low pavement design temperature at the pavement surface is calculated using the equation:

$$T_{pav} = -1.56 + 0.72\,T_{air} - 0.004\,Lat^2 + 6.26\log_{10}(H + 25) - Z(4.4 + 0.52\,\sigma_{air})^{1/2}$$

T_{pav} = Low AC pavement temperature below surface, °C
T_{air} = Low air temperature, °C
Lat = Latitude of the section, degrees
H = Depth to surface, mm
σ_{air} = Standard deviation of the mean low air temperature, °C
Z = From the standard normal distribution table, Z = 2.055 for 98% reliability

The Superpave system allows the designers to use reliability measurements to assign a degree of design risk to the high and low pavement temperatures used in selecting the binder grade. As defined in Superpave, reliability is the percent

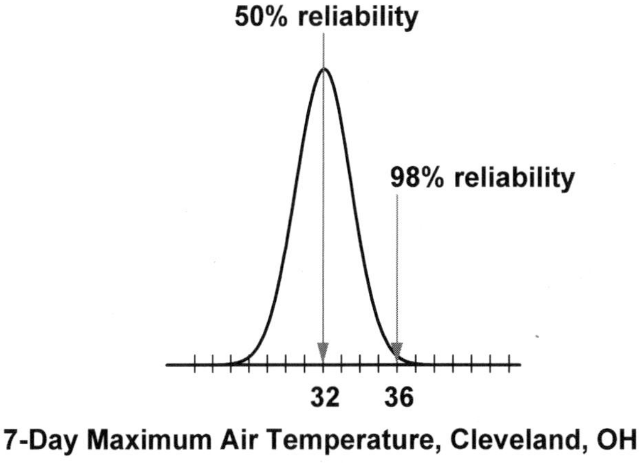

Figure 5.1 Design Temperature Reliability

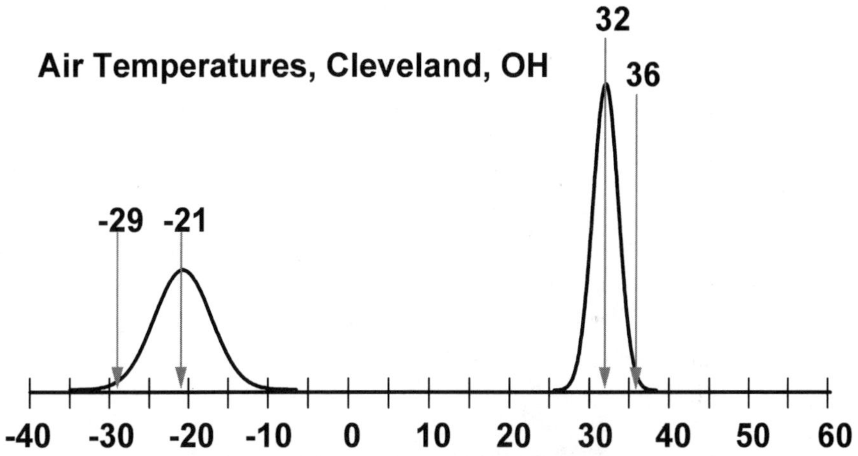

Figure 5.2 Example High and Low Air Temperature Variations

probability in a single year that the actual temperature (one-day low or seven-day average high) will not exceed the design temperatures. A higher reliability means lower risk. For example, consider summer air temperatures in Cleveland, Ohio, which has a mean seven-day maximum of 32°C and a standard deviation of 2°C. In an average year, there is a 50 percent chance that the seven-day maximum air temperature will exceed 32°C. However, assuming a normal statistical frequency distribution, there is only a two percent chance that the seven-day maximum will exceed 36°C (mean plus two standard deviations); therefore, as shown in Figure 5.1, a design air temperature of 36°C will provide 98 percent reliability.

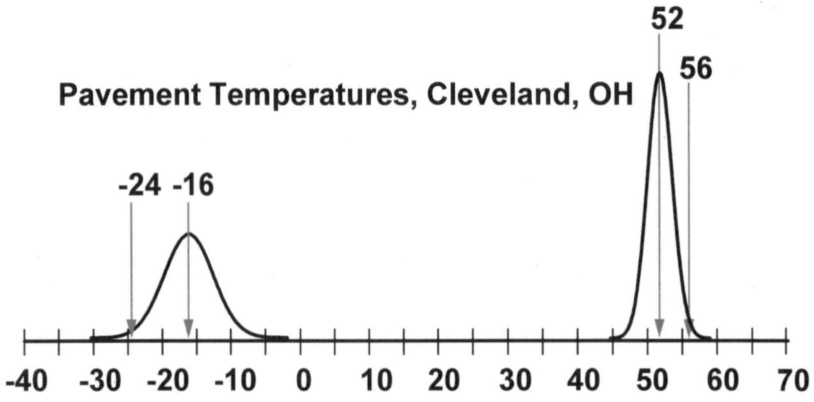

Figure 5.3 Example High and Low Pavement Temperature Variations

Air Temperature Selection

Continuing the example, assume that an asphalt mixture is to be designed for Cleveland. Figure 5.2 graphically represents the statistical variation of the two design air temperatures. In a normal summer, the average seven-day maximum air temperature is 32°C and in a "very hot" summer, this average may reach 36°C. Using a similar approach for winter conditions, Cleveland has an average minimum air temperature of -21°C with a standard deviation of 4°C. Consequently, in an average winter, the coldest temperature is -21°C. For a "very cold" winter, the air temperature may reach -29°C. The standard deviations show there is more variation in the one-day low temperatures than the seven-day average high temperatures.

Pavement Temperature Selection

Continuing the example, for a surface course in Cleveland (latitude = 41.42 degrees) the design pavement temperatures are calculated to be about 52°C and -16°C for 50 percent reliability and about 56°C and -23°C for 98 percent reliability (mean plus two standard deviations). Figure 5.3 graphically represents the statistical variation of the two pavement temperatures.

Binder Grade Selection Based on Pavement Temperatures

To achieve a reliability of at least 50 percent and provide for an average maximum pavement temperature of at least 52°C, the high temperature grade happens to match the design temperature, PG 52. Using the same reasoning, the low temperature grade is a PG -16 to attain 50 percent reliability. Coincidentally, the low temperature grade again happens to match the design temperature, -16. As shown in Figure 5.4, to obtain at least 98 percent reliability, it is necessary to select a high temperature grade of PG 58 to protect above 56°C and a low temperature grade of PG -28 to protect below -23°C. In both the high and low temperature cases of the PG 58-28 binder grade, the actual reliability exceeds 99 percent because of the "rounding up" caused by the six degree difference between standard grades.

This "rounding up" introduces conservatism into the binder selection process. Another possible source of conservatism occurs when considering the same steps encountered during asphalt binder test classification. Although a specific binder may pass all of the criteria when tested at lower or higher temperatures, it will nevertheless be classified by "rounding down" to the next "six-degree" step of the grading system. The net result is that a significant factor of safety is included in the binder selection scheme. For example, it is possible that the PG 52-16 binder, selected previously for a minimum of 50 percent reliability for Cleveland may actually have been graded as a PG 56-20, had such a grade existed. Users of this temperature-based stepped grading system for binder selection should recognize that considerable safeguards are already included in the process. Because of these factors, it may not be necessary or cost-effective to require indiscriminately high values of reliability or abnormally conservative high or low temperature grades.

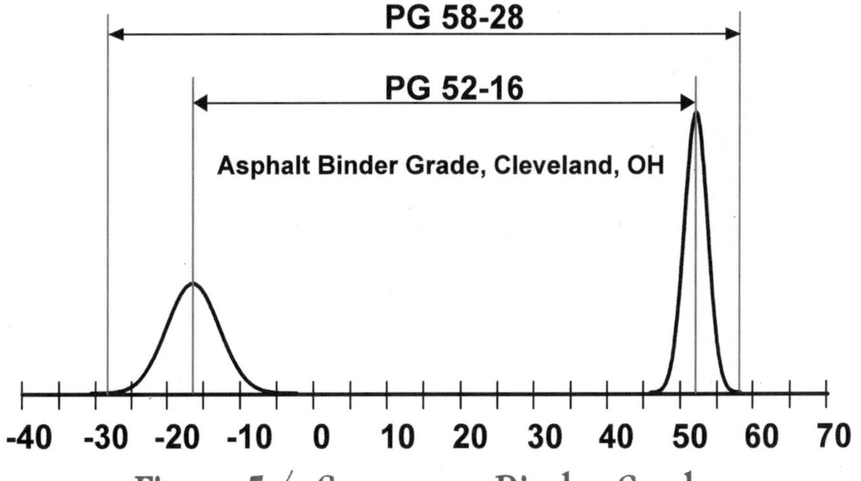

Figure 5.4 Superpave Binder Grade Selections for Cleveland

The Superpave computer program performs all of these calculations based on minimal user input. For any site, the user can enter a minimum reliability and the software will calculate the required asphalt binder grade. Alternatively, the user can specify a desired asphalt binder grade and Superpave will calculate the reliability obtained for the climate at the site. Consequently, agencies are faced with engineering management decisions. They will have to decide the level of reliability to be used. Depending on the policy established by each individual agency, the selected reliability may be a function of road classification, traffic level, cost, and other factors.

It should be emphasized that proper or conservative binder selection does not guarantee total pavement performance. Fatigue cracking performance is greatly affected by the pavement structure and traffic. Permanent deformation or rutting is directly a function of the shear strength of the mix, which is greatly influenced by aggregate properties. Pavement low temperature cracking correlates most significantly to the binder properties. Engineers should try to achieve a balance among the many factors when selecting binders.

Adjusting Binder Grade Selection For Traffic Speed and Loading

The asphalt binder selection procedure described is the basic procedure for typical highway loading conditions. Under these conditions, it is assumed that the pavement is subjected to a design number of fast, transient loads. For the high temperature design situation, controlled by specified properties relating to permanent deformation, the speed of loading has an additional effect on performance. AASHTO MP-2, ***Standard Specification for Superpave Volumetric Mix Design***, includes an additional shift in the selected high temperature binder grade to account for slow transient and standing load applications. Similar to the time-temperature shift described with the test temperature for the BBR (testing at 10°C higher temperature reduced the test duration from 2 hours to 60 seconds), higher maximum temperature grades are used to offset the effect of the slower loading speed. For slow moving design loads, the binder should be selected one high temperature grade to the right (one grade "warmer"), such as a PG 64 instead of a PG 58. For standing design loads, the binder should be selected two high temperature grades to the right or two grades "warmer", such as a PG 70 instead of a PG 58.

Also, an additional shift is proposed for extraordinarily high numbers of heavy traffic loads. These are locations where the design lane traffic is expected to exceed 10,000,000 equivalent single axle loads (ESAL). An ESAL is defined as one 18,000-pound (80-kN) four-tired dual axle and is the unit used by most pavement thickness design procedures to quantify the various types of axle loadings into a single design traffic number. If the design traffic is expected to be between 10,000,000 and 30,000,000 ESAL, then the engineer may consider selecting one high temperature binder grade higher than the selection based on climate. If the design traffic is expected to exceed 30,000,000 ESAL, then the binder should be selected one high temperature grade higher.

References

Additional, more detailed information regarding the SHRP research and the Superpave binder and mixture specifications can be found in these SHRP publications. These publications include:

"The Superpave Mix Design Manual for New Construction and Overlays," Report SHRP-A-407.

"The Superpave Mix Design System - Manual of Specifications, Test Methods, and Practices," Report SHRP-A-379.

"Binder Characterization and Evaluation - Volume 3: Physical Charaterization," Report SHRP-A-369.